# Molecular Biology Biochemistry and Biophysics

# 10

R. J. Lukens

# Chemistry of Fungicidal Action

With 8 Figures

Springer-Verlag New York · Heidelberg · Berlin 1971

*Raymond J. Lukens, Ph. D.*

Plant Pathologist

Department of Plant Pathology and Botany

The Connecticut Agricultural Experiment Station, New Haven, CT, USA

ISBN 0-387-05405-7 Springer-Verlag New York Heidelberg Berlin

ISBN 3-540-05405-7 Springer-Verlag Berlin Heidelberg New York

 Library of Congress Catalog Card Number 70-154308. Printed in Germany. Typesetting and printing: Carl Ritter & Co., Wiesbaden. Bookbinding: Großbuchbinderei Karl Hanke, Düsseldorf

*To my wife*

## Acknowledgments

An author is not an island unto himself. His ideas are formulated from reflections of what he reads, examines and discusses with others. I am indebted to my collegues at The Connecticut Agricultural Experiment Station, New Haven, with whom I discussed many aspects of the book and especially to A. E. DIMOND, H. D. SISLER, J. G. HORSFALL, S. S. PATIL, and M. L. ZUCKER, who took the time and patience to criticize various parts of the monograph. To them go the praises that may come of my efforts. Criticisms that are surely to arise are mine to endure.

I wish to thank MARY CIPOLLINI, BARBARA WOODING, and NANCY, my wife, for the typing and proof reading that they cheerfully performed.

R. J. LUKENS

New Haven, April 1971

# Contents

Chapter 1

# Chemical Control of Deterioration by Fungi

## Introduction

The threat of famine, the potential loss of a wine industry, and the question of what to do about rotting railroad ties were major issues during the 19th century that precipitated investigations into control of the degradative actions of fungi.

In 1845 and 1846, the Irish lost their potato crop because of epidemics of the late blight disease. Little was known about the cause of the disease or ways of controlling it, and the proper means of applying a chemical control were unknown. Lacking their staple food, the peasants starved. The famine was the severest of several economic squeezes in Ireland at the time. As a consequence of the late blight, a million people died and a million and a half emigrated (LARGE, 1940). The catastrophe caused botanists to turn their attention to control measures for plant diseases.

Three decades later, when the downy mildew disease was destroying the vineyards of France, Millardet stumbled upon a cure. The brew of copper sulfate and lime that the farmers applied to roadside rows of grape to discourage pilferage protected the plants from disease. The word was spread, the disease was arrested, and Millardet was a hero. The successful cure stimulated investigations into the uses of chemicals for the control of plant diseases, which hitherto had been haphazardly undertaken. The succeeding years witnessed many successful chemical control measures with materials containing copper, mercury, or sulfur. In addition, an industry for developing and manufacturing equipment to be used in applying these chemicals was begun. The farmer finally had an effective means for minimizing production losses from disease.

The search for industrial fungicides was less dramatic than that for fungicides against plant diseases. Preservatives were found by the route of industrial advancement. Although some preservatives were suggested for treatment of fence posts and ship planks, the railroad industry, developing in the 1800's, set the stage for a serious search for preservatives. How to curtail the rotting of the 3200 wooden ties per mile of track was a tremendous challenge, and a profitable return for a successful treatment was certain. Numerous patents were issued from which proprietary products containing creosote, and salts of mercury, copper, and zinc were developed (MCCALLAN, 1967).

There are many references to chemical control of deterioration before these events, but since the need was not dramatized, the significance of the early work was lost with time, and methods of successful application were never developed. The cleansing power of sulfur was noted by HOMER. HAMBERG considered mercuric chloride as a wood preservative in 1705. A hundred years later, PRÉVOST described the inhibition of germination of smut spores by copper sulfate. These and others are included among the landmarks in fungicide history from 1000 B. C. to the 1950's A. D., listed by HORSFALL (1956).

The search for fungicides that could be used with animals and humans grew out of medicinal investigations and practices to control infectious diseases. Some ointments, powders, and pastes applied to mycotic infections of the skin and mucous membranes to relieve inflammation, irritation, and pain have been demonstrated to possess some fungitoxic properties. Some fungicides suitable for treating animals were borrowed from the plant pathologist; others were selected from original assays of dermatophytic fungi. However, the development of fungicides of highly selective action was provided by discoveries of antifungal antibiotics.

## Importance of Fungicides

Presently, fungicides play an essential role in the production of agricultural crops, in industrial production, and in prolonging the utility of manufactured products. It has been estimated that fungicides are employed in the growing of one-half of the world's crops (ORDISH and MITCHELL, 1967). The production of fungicides, including agricultural and industrial fungicides, in 1965 was 150 million pounds (SHEPARD, MAHAN, and FOWLER, 1966). This amounts to $ 48.6 million and represents about 10% of pesticide sales (SHEPARD, MAHAN, and FOWLER, 1966).

Of the 1,000,000 or so known fungi, fewer than 200 are considered plant pathogens and fewer than a dozen of these cause serious loss (ORDISH and MITCHELL, 1967). However, losses of food, fiber, and ornamental crops from the action of plant pathogens are held to a minimum of 3.3 billion dollars annually in the United States through the use of fungicides (LECLERG, 1964). When systemically acting fungicides become available these losses could be reduced by half.

The increase in food production for the enlarging world population will have to come from intensified agricultural practices because most tillable land is being farmed already. Farming in the temperate zones is intensive, with chemical disease control methods widely employed in the production of food staples. This is not so in the equatorial regions. Undoubtedly, fungicides will play an important role in the intensification of farming in these areas. Since weather, temperature and intensity of sunlight differ in the two climatic regions, such usage will necessitate a reevaluation of fungitoxic materials, their formulation, and means of application.

The health hazards of pesticide usage were dramatized by the publication of "Silent Spring" by RACHEL CARSON (1962). The prerequisites of high fungitoxicity and low mammalian toxicity, especially of materials applied to edible crops, have been paramount in the development of fungicides. Likewise, emphasis was placed on minimizing phytotoxicity and destructive action toward manufactured products by fungitoxic formulations. However, the need to minimize residues and to encourage proper usage of fungicides cannot be overemphasized. A thorough knowledge of the actions of fungicides is imperative to minimize the hazards of usage.

## Actions of Fungicides

Effective usage of fungicides requires a working understanding of just how fungicides perform. Uses of fungicides (McCallan, 1967), their chemical properties (Frear, 1942), how they are applied and how they act (Martin, 1964) have all been the subjects of extensive study. Horsfall (1945) stressed the importance of assay and the significance of the dosage-response curve in interpreting fungitoxic mechanisms. Later (1956), he emphasized the importance of the hydrophobic characteristics for permeation of toxicants. The structure-activity relationships of fungicides and other toxicants were examined by Sexton (1953). Factors regulating selective action were examined by Albert (1960). Recently, Torgeson (1967 and 1969) has edited a comprehensive treatise that covers many facets of application and mechanisms of action of agricultural and industrial fungicides. Chemical control of mycotic infections in animals and man has been discussed by Sternberg and Newcomer (1955) and more recently by Hildick-Smith, Blank and Sarkany (1964).

In the present monograph, data are based upon literature of plant pathology, pharmacology, and industrial preservatives. Fungitoxic chemistry is presented from a mechanistic viewpoint. Activity is described in terms of the physical and chemical properties of the compounds in relation to recent concepts of cellular biology. The principles developed may be applicable to an understanding of the actions of fungicides in the control of diseases of plants, animals, and man as well as the action of fungicides in preventing deterioration of material.

Chapter 2

# Measurement of Fungitoxicity

## Introduction

Fungitoxicity has been defined as the ability of a chemical to interfere in an adverse way with the vital functions of a fungus by physiochemical means (ANON, 1943). Fungistatic action implies a continuous interference as long as the organism is in the presence of the toxicant. Fungicidal action denotes a persistent action after withdrawal of the toxicant. Unfortunately, there is no single site of action on which to base an assay of fungitoxicity. Thus, criteria for toxic action are described in general terms as inhibition of growth and development. Growth, swelling of cells, collapse of protoplasts, sporulation, germination of spores, and respiration all are responses used to determine fungitoxicity. Fungitoxicity can also be expressed in numerical terms as the dosage of toxicant that produces a unit inhibition of fungal response in a prescribed period of time.

This chapter will deal with responses of fungi to toxic agents, rather than methods of measurement or statistical analysis. For more thorough discussions on measurement and statistics, other sources are available (MANDELS, 1965; LILLY and BARNETT, 1951; ANON, 1962; FINNEY, 1964; and BLISS, 1957).

## Fungitoxic Action

### Against Fungous Responses

Fungitoxicity is usually measured in terms of the response of a treated culture relative to that of a control.

*Inhibition of Growth.* Fungicides inhibit growth, and fungous growth can be measured in a variety of ways. On nutrient agar, growth can be measured in terms of colony diameter in Petri dishes, by length of colony in horizontal tubes or by surface area of the colony. In the roll-culture technique (MANTEN, KLÖPPING and VAN DER KERK, 1950), agar media are mixed with fungicide and inoculum in cylinder bottles, which are slowly rolled until the agar solidifies. After incubation, the amount of growth is graded according to one of 5 established categories.

At times, growth inhibition is measured in terms of reduction in dry weight. However, the harvesting of fungus may be difficult and may cause error in measurement. On heating agar to recover the fungus, appreciable loss of water-soluble and volatile constituents is likely to occur. Often, components of the media and insoluble fungicides may be difficult to eliminate from the fungus. In situations where fats accumulate instead of new protoplasm, dry weight may be misleading.

With bacteria, yeasts, and other unicellular organisms, dry weight can be equated with cell volume, cell number, and turbidity. In the cell volume and number methods, small aliquots can be withdrawn from cultures for measurement without disturbing the entire culture. Such a feature is desirable when one is studying growth rates. Turbidity measurements can be taken of a whole culture with only slight disturbance to the culture or of small aliquots which are removed from the culture for reading and then discarded. The failure to distinguish between cell division and swelling is a deficiency of the cell volume and turbidity criteria. Certain toxicants may inhibit division of cells but permit cells to increase in size. In such cases, other measurements of growth should be sought.

In addition to dry weight, protein synthesis can be used to measure growth of fungi for toxicity measurements. Total protein content of a fungus is related to growth and does not measure storage products of the fungus. This technique has not been popular in fungitoxic studies.

*Inhibition of Spore Germination.* Most fungicides are designed to prevent establishment of the causal fungus. Spores are the principal inoculum for establishment. Thus, methods to measure viability of spores are of prime interest to plant pathologists who are anxious to develop control measures for disease. The development of the many fungicides between 1945 and 1960 was made possible by the spore germination test as described by McCallan (1930). In essence, a small volume of water containing spores is placed on microscopic depression slides containing residues of the toxicant. To ensure high viability, orange juice, vitamins, or buffer are added to the spore suspension. A valid test necessitates less than 5% ungerminated spores in untreated cultures to ensure that lack of spore germination is caused by fungitoxic action.

At times, mycologists, by necessity, are forced to work with an organism that germinates poorly. In these cases, other tests for viability have been developed. With powdery mildew spores, Koopmans (1959) observed that viable spores swell in a humid atmosphere and dead ones fail to do so. Rather than look for germ tubes, Koopmans looked for enlarged spores. Hashioka and Takamura (1956) measured response of spores of *Cochliobolus miyabeanus* to solutions of mercurial compounds by recording percentage of spores with coagulated protoplasts. Collapsed conidiophores or coagulation of their protoplasts is one criterion used by Horsfall and Lukens [1966 (2)] to determine viability in their sporulation test with *Alternaria solani.*

*Inhibition of Respiration.* Respiratory studies are necessary for mechanistic studies of fungitoxic action, but they are useless as a screen for fungicides unless one is looking for respiratory inhibitors. McCallan, Miller and Weed (1954) found a wide variation between inhibition of respiration and inhibition of spore germination of fungicides. Generally, spore germination was more sensitive. Obviously, fungicides attacking respiratory enzymes are strong inhibitors of respiration. Others may reduce respiration, indirectly, by attacking sites close to the respiratory tract. And still others, which inhibit cell division or protein synthesis, may have little detectable effect on respiration within the normal period of observation.

*Inhibition of Sporulation.* An antisporulant is a chemical that prevents reproduction by spores. The poison food technique, frequently employed in studying antisporulant action, fails to distinguish effects on sporulation from those on growth. Horsfall and LUKENS [1966 (2)] have minimized the growth factors with *Alternaria solani* by treating conidiophores prior to induction of sporulation. Sporulation, growth, or collapse of conidiophores is recorded at the end of the spore induction period. SHEPARD and MANDRYK (1964) examined antisporulant action of infections of *Peronospora tabacina* in tobacco leaves. After discs of diseased leaves are floated on a solution of test chemical, spore yield is determined by counts made with the aid of a hemocytometer.

### Against Fungous Decay

In microbial deterioration studies, the soil burial test has been widely used to screen fungicides. In this test, treated cloth or stakes are buried in soil containing known degrading fungi or a standard spawn for varying periods of time, after which they are removed and graded for deterioration. Data are recorded either on the amount of decay per unit of time or on the time required to obtain a certain amount of decay. With fabrics, tensile strength of the cotton fiber has been a popular criterion of measure for deterioration.

In studies with plant diseases, fungitoxicity has been measured by the reduction in amount or in rate of development of diseases. A disease assay may be as simple as using discs of cucumber leaves dusted with spores of powdery mildew or as extensive as using rows of apple trees subjected to natural inoculum of *Venturia inequalis*. Many assays fall between these two extremes. The elaborate greenhouse trials of *Alternaria solani* on tomato plants (WELLMAN and MCCALLAN, 1943) and *V. inequalis* on apple seedlings (HAMILTON, 1958) are two noted examples.

A popular measure of disease severity is based upon 5 or 6 equally spaced grades that range from no disease to complete destruction of tissue. The Horsfall-Barratt system (1945) is based upon the ability of the eye to distinguish logarithmic differences accurately. Their 12 grades ranging from 0 to 11 represent percentages of disease — 0, 0 to 3, 3 to 6, 6 to 12, 12 to 25, 25 to 50, 50 to 75, 75 to 87, 87 to 93, 93 to 96, 96 to 100, 100, respectively. Evaluation of seed treatment and treatment for control of damping off can be based on direct counting of emerged seedlings.

Laboratory assays of fungitoxicity together with accessory tests that measure deposition and persistence of toxic residues can be used with considerable accuracy to predict field performance of a candidate fungicide (RICH, HORSFALL and KEIL, 1953). If the variables that act on a toxic residue in the field are known, they can be tested separately in simple laboratory tests. The trend has bypassed individual laboratory tests, however, and has swung to miniature tests under somewhat controlled conditions that mimic those in the field. The assumption is that one integrated test can distinguish potential candidate compounds. In addition to protective action, the miniature tests on plant disease can detect systemic action. However, because such tests require days or weeks instead of hours, the rate at which compounds can be evaluated is sharply reduced as compared with laboratory tests.

Often, a likely candidate fungicide developed from laboratory and greenhouse tests fails in the field. A probable failure in the testing procedure lies in the technique of inoculating with aqueous suspensions of spores. When spores are sprayed on the treated surface, the water provides a contact between spore and particle of fungicide. However, in nature, spores of some fungi can germinate and infect plants in the absence of free water. Thus, a successful fungicide in practice must contact these fungi in other ways; *e.g.*, by vaporizing or by partitioning into the waxy cuticle of the leaf and attacking the fungus as it penetrates this layer. The water artifact in testing procedures can be avoided by applying dry deposits of spores to the treated surface. A vacuum duster was developed by the Crop Protection Institute to inoculate bean plants with dry rust spores.

In the search for chemotherapeutants for controlling mycotic infections of animal and man, fungous response at the infection site must be assayed precisely to determine the successful movement of a fungicide from point of application to site of infection and to inhibition of the pathogen. The fungus consisting of a high amount of growth in its natural habitat may respond differently to the fungicide than when it is inoculated in small amounts on poisoned medium. However, precise fungitoxic data from clinical assays are difficult to obtain. Difficulty arises from the variability of response of test animal. The use of inbred lines and germ-free animals may overcome some of the variability (Czerkowicz and Stuart, 1968).

## Interpretation of Fungitoxicity

Most measures of fungitoxicity fall into one of two categories: (1) the all-or-none response, as determined indirectly in spore germination tests, and (2) the graded response as determined in growth and respiration tests. How much information is obtained from either depends upon the variation in the fungus and the relation between dosage and response. A standard fungicide included in experimental runs can serve as an indicator of fungous variability and dosages are then chosen in the range between zero and complete responses.

### Dosage-Response Curves

Linear dosage-response curves are required to evaluate the fungitoxic characteristics of chemicals. However, raw data on fungitoxicity need adjusting to obtain linear curves. The dosage is expressed logarithmically to remove the *j*-shaped component of the curves, which Horsfall (1945) attributed to the law of diminishing returns. That is, each succeeding increment of dosage causes a smaller unit of fungous response. Thus, biological responses change arithmetically with logarithmic change in stimulating agent. When data of graded responses are plotted against log-dosage, an extended sigmoid curve is formed (Bliss, 1957). The long central section of the curve can be used as the linear expression for that curve.

However, the plotting of percent inhibition of spore germination against log-dosage gives sigmoid curves with nonlinear central portions. The sigmoid curve, common to many biological phenomena, is a variation of the bell-shaped frequency-distribution curves for sensitivity of fungous cells to a toxicant. Bliss (1957) defined the symmetrical sigmoid curve as a cumulative normal distribution. That is, a few individual spores are extremely sensitive or resistant to the toxicant, but the majority fall in the middle. Thus, each increment in dosage of fungicide gives unequal increments of affected individual spores. The percentage of spores in a culture indicates the probability that one spore will be inhibited. Gaddum (1933) introduced his normal equivalent deviation for use in all-or-none assays. Bliss (1935) avoided the negative values of GADDUM's units by adding 5 to these values and called his units probits. He (1957) concluded that the plotted points can be fitted adequately to a straight line if the data define a normal distribution of the log-threshold dosage of individual spores. Sources for nonlinearity of the probit log-dosage curve lie in the manipulative technique, in the mode of action of the toxicant or in the dosage actually reaching the organism (HORSFALL, 1945). Percent inhibition of sporulation can be plotted directly on log-probability graph paper to obtain fungous responses in probit form and dosage in logarithms. Probits were first applied to spore germination data by WILCOXON and McCALLAN (1939) and to field data of disease by DIMOND *et al.* (1941). A statistical treatment of probit analysis has been written by FINNEY (1964).

The WEIBULL (1951) distribution was adapted by COOKE (1966) to measure the probability of disease control by an effective deposit of fungicide on plant foliage. The log of the WEIBULL distribution, log (ln 1/1-P), is plotted against the log of effective deposit of toxicant, log $(x - x_u)$, to give a linear response, where P = probability of control, x = concentration of fungicide and $x_u$ = threshold concentration of fungicide.

*Index of Fungitoxicity.* A dosage-response curve contains information from which an index of fungitoxicity can be derived. The dosage of a fungicide that gives a desired equivalent response, *e.g.* threshold, ED-95 or ED-50 value, is commonly used to index potency. Although the threshold dosage — the lowest dosage to give 100% kill of a population — has practical value in chemical disinfestation, a definition of a minimum lethal dosage is unattainable when applied to a population. Individual cells vary in susceptibility and the emphasis is placed upon the most resistant cells. The ED-95 value is a wiser choice than the threshold value. ED-95 has a practical relationship to chemical control and is not distorted by the few extremely resistant cells. Although the ED-50 is less valuable than ED-95 to chemical control of fungous decay and disease, ED-50 is popular in laboratory assay for indexing efficacy. It requires less work in the laboratory to obtain reliable values for ED-50 than it does for ED-95.

However, ED-50 values are strongly influenced by the fungus, conditions of assay, and the slope of the dosage-response curve. Fungous species differ in their sensitivities to a toxicant. Susceptibility has been related to spore volume, shape, exposed surface, cell and nuclei number (DIMOND *et al.*, 1941). Susceptibility can be related to the capacity of the fungous cell to enter chemical and physical reactions with the toxicant. This will be elaborated upon in later chapters.

Age may increase or decrease sensitivity of fungous cells to toxicants (OSTER, 1934; DIMOND and DUGGER, 1941; DIMOND *et al.*, 1941). Factors that change with senescence, such as nonfunctional budscars of mother cells of yeast (MORTIMER and JOHNSTON, 1959), apparently changed the susceptibility of *Saccharomyces pastorianus* to captan between tests conducted on successive days of the week. ED-50 values for tests on Monday through Thursday decreased 0.35, 0.29, 0.27, 0.25 μg/mg dry weight, respectively (LUKENS, unpublished data). Although inoculum for each test was from a 24 h culture, the cultures were transferred daily except Saturday. Thus, the population of tough old mother cells would be highest on Monday and lowest on Thursday.

ED-50 values are proportional to spore load on a log-log basis (DIMOND *et al.*, 1941; MCCALLAN, WELLMAN and WILCOXON, 1941). The spore load of 50 spores per field at 100 × magnification (used by many workers) is a compromise to simplify the recording of data and to minimize the error from spore load.

Environmental factors that have been reported to affect ED-50 values are temperature, pH of medium, and nutrient content of medium. ED-50 values decrease when temperature is lowered or raised from the optimum for spore germination and growth (DIMOND *et al.*, 1941; THORNBERRY, 1950). A drop in pH of medium from 7.5 to 6.0 to 4.5 increased the potency of captan to *Saccharomyces pastorianus* 15 × and 6 × in the two respective intervals and growth increased 5 × and 2 ×, respectively [LUKENS and SISLER, 1958 (1)]. Each unit increase in growth by pH manipulation produced a 3-fold increase in susceptibility. In both filamentous growth and spore germination tests, enrichment of medium can increase, decrease, or not affect ED-50 values (DIMOND *et al.*, 1941; MANDELS and DARBY, 1953; MILLER, 1950; MCCALLAN and WILCOXON, 1939).

*Slope.* The slope of the dosage-response curve is the ratio of change in fungous response to the corresponding change in dosage. Slopes measure potencies under conditions of changing dosages. More precisely, in spore germination tests, slope is a probit of inhibition per log unit of dosage (BLISS, 1935). DIMOND *et al.* (1941) introduced slope as a measure of potency to complement that of ED-50 values.

The steepness of a slope is influenced by the fungus, by the conditions of the experiment and by the fungicide. Slopes of dosage-response curves of a fungicide vary with the test organism. With the same organism, young spores produce steeper slopes than old spores (DIMOND *et al.*, 1941). However, the level of inoculum does not affect slope. Nutrient (DIMOND *et al.*, 1941), pH of medium [LUKENS and SISLER, 1958 (1)] or any factor that favors spore germination and growth tends to promote steep slopes in dosage-response curves. Apparently, slopes are steeper when fungous cells respond uniformly under conditions for maximum germination and growth.

Conversely, the diversity of conditions from the optimum or an increase in a number of limiting factors in the experimental design tends to flatten slopes of dosage-response curves. Slopes for chemical control of disease in the field are flatter than slopes for inhibition of spore germination in the laboratory (DIMOND *et al.*, 1941). Slopes are flattened by improving coverage of toxicant over the treated surface or by weather or any factor that discourages disease. Also, slopes are flattened or decreased with reduction in incubation time.

Since slope measures potency with changing concentration of fungicide, slope is a characteristic of the fungicide and may vary with the fungicide.

Slope affects the index of toxicity, and the extent of that effect depends upon the level of equivalent response used in the indexing. DIMOND *et al.* (1941) found that dosage-response curves of a toxicant tend to converge toward 100% inhibition when the slopes are intentionally altered by manipulating factors that affect slope. Thus, the ED-50 value varies widely with slope, the ED-95 value varies slightly with slope and the threshold dosage for 100% inhibition (in theory) is not affected by slope. To reduce the effect of slope on ED-50 values, conditions of assay must be precisely controlled.

Slopes qualify the comparisons of ED-50 values between toxicants. In the bioassay of toxicants in samples of unknown quantities, precise comparisons between ED-50 values necessitate parallel dosage-response curves (BLISS, 1935). The equal slope prerequisite insures that the differences in ED-50 values between known and unknown are accurate measures of the difference in concentration of toxicants in causing toxicity.

However, on indexing the potency of chemicals that have different slopes, the parallel slope prerequisite is discarded. ED-50 values are the rule. The effect of slope is small when ED-50 values are widely scattered. When ED-50 values are close together, factors other than potency become decisive in the choice of fungicides.

Slopes from spore germination data may describe, in part, the mechanism of action of fungicides (HORSFALL, 1956). Equal slopes suggest that the products of fungous responses to the toxicants are equal. Since structurally related compounds may have common factors of action, equal slopes may describe the same mechanism of action and unequal slopes may describe different mechanisms of action. However, with structurally dissimilar compounds where factors of toxicity vary significantly, equal slopes do not necessarily describe the same mechanism of action.

## Criteria of Dosage

As long as fungicides are with us, we will be confronted with the problem of defining a basis of universal acceptance for dosage. Dosage means, to the farmer, how much material to mix in his spray tank for applying to plants to prevent disease. To the scientist, dosage presents the problem of finding a basis which gives meaning to the ED-50 values used to compare fungicides. The problem lies in defining the amount of fungicide needed to render a unit kill of fungi. The nagging part lies in the definition of kill. Thus, the basis for dosage varies with the individual investigator.

Normally, dosages are based upon the applied concentration of fungicide. They are often expressed in terms of the concentration of chemical in the external medium. Where protective fungicides are of primary concern, amount of toxicant per unit area of surface is a meaningful way of expressing dosage (HORSFALL, 1956). When the amount of inoculum is known, dosage can be easily converted to amount per unit of organism. ED-50 values expressed in terms of concentration in external medium are useful for comparing fungicides for practical purposes. However, in the laboratory, ED-50 values in terms of applied dosage per unit weight of organism are required for comparison between fungi.

Dosages expressed in weight of fungicide are all that is available for compounds with unknown formula weights, as in the case of many antibiotics. The weight basis is commonly used in studies of structure and activity where differences in molecular weights among analogous compounds may be small in contrast to the fungous response to the compounds. However, for more precise measurements of fungitoxicity in studies on mechanism of action, applied dosages of compounds can be expressed on molarity.

Dosage based upon the amount taken from the medium by fungous cells has been suggested to measure the real toxic dosage of fungicide received by the fungus (MILLER, MCCALLAN and WEED, 1953; MCCALLAN, BURCHFIELD, and MILLER, 1959).

MARTIN (1964) agrees with the internal dosage concept and suggests that toxic dosage expressed in terms of the external concentration may be misleading. He bases his argument on the work of FERGESON and PIRIE (1948), who found that toxicity of chemical reactants fails to follow thermodynamic equilibrium phenomena in the external environment, thus differing from toxicity of physical toxicants. Physical toxicants disrupt cellular structure on a gross scale, and toxicity of compounds in a series is related to the thermodynamic activities of the compounds. Toxicity occurs near the equilibrium of the thermodynamic activity. In contrast, chemical toxicants are active at low thermodynamic activities (less than 10% of maximum) where equilibria are not obtained.

However, the chemically reactive fungicides captan, dyrene, and glyodin are toxic at high thermodynamic activities (BYRDE and WOODCOCK, 1959). These fungicides are poorly soluble in water and readily enter reactions of detoxication. Furthermore, the high activity coefficients of compounds with extremely low water solubility (HIGUCHI, 1960) can accentuate the nontoxic reactions.

With uptake dosages, one is led to assume that the fungicide absorbed by the fungous cell is available for toxic reactions. However, many pitfalls lie in this line of reasoning, which will be discussed in Chapter 4. Excessive degradation of fungicide during uptake by the cell reduces the amount of material available to sites of toxicity. The failure to appreciate the rapid degradation of captan by fungous cells caused OWENS and NOVOTNY (1959) to assume erroneously that captan removed from the external solution by conidia of *Neurospora* would be available for toxicity for a period of time as determined by the hydrolysis rate of the fungicide. By basing ED-50 values on uptake data, RICHMOND and SOMERS (1963) concluded that uptake of captan by conidia of various fungi had no relation to toxicity in spite of every dosage-response curve of the fungicide testifying to the contrary. Later (1966) they clarified their confusion by verifying that much of the captan taken up by fungous cells is degraded by cellular thiols. The discrepancies between uptake and external dosage curves for a series of *s*-triazines reported by MCCALLAN, BURCHFIELD and MILLER (1959) may be caused by the differences in the lipophilic and chemical properties of the compounds. Undoubtedly, these properties are involved in permeation as well as in reactions of the toxicants with cellular constituents.

At the ED-50 value, apparently, all of the toxicant available to the fungus from the external solution is exhausted. A reserve of captan and dyrene in the medium is required to cause 100% inhibition of spore germination (OWENS and NOVOTNY, 1959, and BURCHFIELD and STORRS, 1957, respectively). Dosage based upon uptake does not include the portion of dosage in the medium necessary for toxicity.

## Conclusions

Inhibition of growth, spore germination, respiration, fungous deterioration, and disease development are used to measure fungitoxicity. Many factors of performance can be examined in simple laboratory experiments to reduce the waste of failures in the field. Dosage-response phenomena are described in forms suitable for obtaining linear curves for computing indexes of fungitoxicity (ED-50, ED-95 or threshold dosage). The slope of the curve reveals the dynamics of toxic action. Slopes of curves for several fungicides must be parallel for statistical comparisons of ED-50 values. However, toxicants displaying different slopes can be compared for practical purposes by their ED-50 values. An integration of ED-50 and slope has not been found for precise comparisons of fungitoxicities of compounds with different slopes.

The basis for dosage varies with the object of study. Concentration of toxicant in the applied mixture and weight per unit area are employed in performance studies on protecting surfaces. Weight or molarity of applied dosages per unit weight of fungus is often used in laboratory studies to index fungitoxicity and is an important basis for dosage in mechanistic studies where toxic action is limited by the thermodynamic activity of the compounds. Uptake dosage per unit weight of fungus is essential to examine mechanisms of action, but this measure may lead to erroneous conclusions for toxicants that decompose and are detoxified by the fungus.

Chapter 3

# Fungitoxic Barriers

## Introduction

Chemical control of fungous growth depends upon the performance of the fungicide when it is deposited in small amounts on the treated surface. The regulation of fungicidal action depends upon the use of the material to be protected. With agricultural products, the fungicide should be effective for a limited period of time. When mixed with soil, the residues must be nontoxic to plants at the time of planting. Edible crops must be free from fungicide at harvest. However, when applied to textiles, wood, and manufactured products, the fungicide should prevent deterioration during the life of the product. To control mycotic infections of animal and man, the fungicide must be nontoxic to the host. Thus, an understanding of the underlying factors is imperative in order to restrict performance to the time required.

The performance of a fungicide is governed by its deposition and retention on the surface. Factors affecting these processes lie in the fungicide, its means of application, and the nature of the surface to be treated. The action of fungitoxic residues will be discussed in terms of their application, deposition, and retention on the treated surface.

## Methods of Application

### Animal Fungicides

Chemical control of fungi attacking animals and man necessitates several means of application. Living quarters, articles of clothing, and objects of contact are treated with disinfectants to eliminate sources of fungous inoculum. Hair, skin, and other exterior sites of infection are treated with powders, ointments, and pastes containing fungicides to arrest the fungus at the sites of infection as well as to allay the irritation, inflammation, and discomfort from infection. Infected mucous membranes are bathed with solutions containing fungicides and intestinal infections are treated with oral preparations. A few antifungal antibiotics that act systemically and cause no side effects in the host are administered orally if they can pass through the intestinal wall or are administered intravenously if they cannot.

The common disinfectants, quaternary ammonium surfactants, chlorine oxidants, and formaldehyde are highly effective fungicides. Fungitoxic fatty acids, 8-hydroxyquinoline, benzothiazoles and quinones that are nonirritating to the host have been employed as topical antiseptics to combat dermatophytic infections of *Microsporum*, *Trichophyton*, and *Epidermophyton*. Derivatives of silver, mercury, and zinc are effective fungicides for mycotic infections, also. Potassium permanganate, aluminum acetate and boric acid, although mildly fungitoxic, allay symptoms of inflammation and irritation (Oster and Woodside, 1968).

Several antifungal antibiotics have been proven highly successful for topical and systemic treatments of mycotic infections. Griseofulvin administered orally or topically is effective against the dermatophytic fungi (Gentles, 1958). Several polyene antibiotics are effective, but most are restricted to topical application. Nystatin, a tetraene, and trichomycin, a heptaene, inhibit growth of *Candida albicans* in mucous membranes and intestines (Oster and Woodside, 1968). Pimaricin inhibits growth of *Candida vagembis*. Amphotericin B having systemic activity can be administered intravenously. The antibiotic X-5079 C, a polypeptide from *Streptomyces*, is toxic to many fungi causing mycosis (Emmons, 1961).

Excluding the sanitary fungicides, chemical control of mycotic infections of animals and man differs markedly from that of plant disease. Firstly, the skin of an animal presents a protein as well as a lipoid barrier to penetration of fungicide from the surface to a subcutaneous infection site, while the cuticle of plants consists largely of wax. Secondly, systemic movement through the animal is by way of a closed circulatory system. Plants have two contrasting systems for passage of chemicals, and each is open-ended. In the xylem chemicals move with the transpirational stream from roots to leaf tips and in the phloem most chemicals move with the mass flow of sugar. Thirdly, the strategy of chemical control is therapy or eradicating the pathogen in the infection. In contrast, strategy of control in plant diseases is primarily prevention or protecting the plant from fungous invasion. Since the toxic dosage of a fungicide depends upon the amount of fungus, its vitality and source of food, arresting growth of an established pathogen at the site of infection contrasts sharply with preventing establishment of the pathogen on the surface of the host. Techniques of topical, oral, or intravenous application of a toxicant lie beyond the competence of the author and authorities in the medical field should be consulted (Götz, 1967; Sanders and Nelson, 1962; Sternberg and Newcomer, 1955; and Utz, 1967).

## Agricultural Fungicides

Fungicides are applied to soil, seed and propagating material, growing plants, and produce for combating plant pathogens. The method of application and choice of fungicide are peculiar to the crop and the surface to be protected.

*Mixing with Soil, Seed, and Produce.* The primary concern is the placement of fungicide where its toxicity to the pathogen will be most effective. Generally, fungicides are applied to a point source and are dispersed by various means to the pathogens. To rid cultivated soil of pathogens, fumigants and nonvolatile fungicides are mixed in soil, but each in a different way. Fumigants are applied to soil by trench and multiple point methods and the toxicant attacks the pathogen in the vapor state (Page, 1963). The liquid fumigant is deposited to a depth of 6 to 8 in. and spaces of 6 to 9 in. The soil surface is closed immediately after treatment. Halogenated propanes, propenes, and ethanes, methyl isothiocyanate, and sodium methyl dithiocarbamate are typical of fungicides employed as liquid fumigants in the field. They are general sterilants that kill fungi, bacteria, and nematodes in soil. Because of their phytotoxicity, the liquid fumigants are employed at a time prior to planting which is sufficient for the residual vapors to dissipate from the soil.

The vaporous fumigant methyl bromide can be used on small plots, provided a plastic cover is sealed over the soil (SHURTLEFF *et al.*, 1957). The covering is held above the soil surface to allow circulation of vapor from evaporating pans, which are placed approximately 10 ft apart. Methyl bromide, as well as formaldehyde, is commonly used for treating bulk soil for potting and bed use. For best results the soil moisture should be favorable for seed germination and soil temperature above 50 °F (NEWHALL and LEAR, 1948).

Nonvolatile fungicides are drilled and sprayed to the soil surface and then are mixed in soil to a prescribed depth with use of disc or cultivator (PURDY, 1967). MENZIES (1957) has used the broadcast method of application to control infections of *Rhizoctonia* and *Streptomyces* on potato with pentachloronitrobenzene. Stem canker occurring near the soil surface is controlled by 10 to 20 lb per acre, while tuber infections below the soil surface require 40 to 50 lb per acre for control.

Since many pathogens attack row crops only in the seedling stage, the soil adjacent to the planting row is all that must be treated. Moreover, fungicides at nonphytotoxic dosages can be applied to soil during the planting operation. Successful use of dust, wettable powder, granular, and spray formulations has been reported (PURDY, 1967). Damping off of cotton by *Rhizoctonia solani* has been controlled successfully with pentachloronitrobenzene and captan, while damping off by *Pythium* required Dexon.

Certain soil-borne pathogens can be controlled by applying a fungicide to the soil surface with no mechanical mixing with soil. A common practice to control snow moulds of bentgrass turf is the application to turf of mercurous chloride mixed with dressing or fertilizer (MONTEITH and DAHL, 1932). The same type of treatment to wheat controls crown rotting fungi in that crop. With wheat, a talc mixture of fungicide applied by airplane seems more practical to overcome difficulties in the special preparation of fertilizer and fungicide (PURDY, 1967).

Aqueous suspensions and solutions of fungicides have been applied with sufficient water to wet the top inch of soil for the control of damping-off disease in the nursery and greenhouse (VAARTAJA, 1964). Methyl mercury dicyandiamide, captan, pentachloronitrobenzene, thiram and other fungicides are used to control seedling diseases of numerous crops. A single drench of fungicide to turf is as effective as five sprays applied weekly to protect Kentucky bluegrass from *Helminthosporium vagans* Dresch (LUKENS, 1965). Drenches of nabam and 8-hydroxyquinoline sulfate control root infections of turf grasses by *Rhizoctonia solani* (LUKENS and STODDARD, 1961). However, the control of root infections was improved by injecting the fungicides beneath the sod at 200 p.s.i.

Fungicides are mixed with seed to disinfest the seed coat of pathogens and to protect the seedlings from damping-off diseases. PURDY (1967) has reviewed the various methods of application. Dry formulations of fungicide are mixed with seed by hand, a rotating drum or continuous-flow dust treaters. However, dry formulations create a dust and health problem in large-scale operations and limit the dosage of fungicide by the amount of powder that can stick to the seed.

The liquid method of application lacks the disadvantage of dry applications and lends itself to accurate control of dosage. In the slurry method, in which a thick suspension of fungicide is mixed with seed in a continuous operation by machine, both fungicide and seed are metered independently into the treating machine (PURDY, 1967). With fluid formulations, seeds are treated by one of several methods: *e.g.* immersed in solution, wet with small amounts of solution in revolving drums, or treated with mist or spray.

Plant propagating material other than seed is often treated with fungicide before planting. Dipping or immersing fleshy bulbs and potato seed pieces in aqueous formulations of various fungicides has been reportedly successful (PURDY, 1967).

The extension of shelf-life of fresh produce by modern trends in processing and marketing has intensified the economic loss from plant diseases after harvest. Losses are minimized by preventing infection before, during, and after harvest. The avoidance of bruising and wounding produce during harvest reduces the chance for fungous attack after harvest. Treating the produce after harvest with fungicides minimizes the inoculum of the pathogens and protects produce from subsequent inoculations.

Methods of applying fungicides to produce after harvest have been reviewed by ECKERT (1967). In brief, the type of treatment is incorporated into the processing scheme and little residue remains shortly thereafter. Produce may be treated during processing, in storage, and en route to the retail market. The fungicides are chosen for their ability to concentrate at the infection sites or wounds on the produce. Fumigation is the chief means of treating produce. The polar fumigants — sulfur dioxide, ammonia, and low molecular weight amines — being water soluble, are differentially absorbed by the wounds. *Rhizopus*, *Botrytis*, *Penicillium*, and *Monilinia* are controlled by fumigating highly perishable fruits and berries.

Aqueous preparations of fungicides are employed where produce is normally washed or hydrocooled during processing. Solutions are preferred over suspensions because of the uniformity of treatment. Hence, sodium hypochlorite has been added to the water to disinfest stone fruit during the hydrocooling process as well as to prevent microbial growth in the hydrocooler. Citrus fruits have been treated with sodium *o*-phenylphenate in solution, in immersion tanks, and in solutions and foams during the scrubbing operation (ECKERT, 1967). Although good sanitation in storage areas, containers, and wrappings reduces inoculum in transit, fungicides impregnated into the wrappings and containers aid these barriers in warding off invading fungi.

*Applying to Crops.* Although the major portion of agricultural fungicides is applied to crops, the quantity per application is small (about one pound of active ingredient per acre for most crops to seven pounds per acre for putting green turf). The active ingredient is diluted with inert materials in quantities to 300 gallons per acre (FULTON, 1965). The fungicide is actually applied to plants in particulate form as particles of dust or droplets of spray. The particles are produced and are propelled toward the plant by various types of equipment. The types of applicators and their performances have been reviewed by POTTS (1958) and RIPPER (1955).

With dry formulations, the dust is milled to particles 1 to 5 microns in diameter and diluted to about 5% with talc or a similar clay (COURSHEE, 1967). The dust is blown into the air and carried to the plant in the air currents. The rate of application is governed by metering the dust into the air stream, the speed of air leaving the duster, and the speed of the duster moving on the ground. The dusting usually is done in early morning or evening when the air is still, to prolong the time the dusty air surrounds the plant. In addition, a leaf wet with dew in the morning or evening captures more dust particles from the air than a dry leaf of mid-day. Sulfur, captan, zineb, dichlone, maneb, and the fixed coppers are commonly applied as dusts.

Dusts are injected into the air by heat to be carried to crops growing in greenhouses. The firing material either is mixed with the dust and ignited in small containers or is on a wick, which is ignited and thrust into the container of dust. Dosage is regulated by the formulation, the size of the containers and their spacing in the greenhouse. The greenhouse is closed for at least four hours to allow diffusion of dust in the air and its settling onto plants.

The spray nozzle breaks up wet formulations into droplets and propels and distributes them Droplets are formed by one of two mechanisms: (1) the interaction of air and liquid films and (2) internal disruption of forces (COURSHEE, 1967).

Pressure nozzles work by the first mechanism with liquid as the moving phase. In the swirl nozzle, films are formed by the liquid swirling within the nozzle and leaving through a round aperture shaped in a hollow rotating cylinder. In fan nozzles, the liquid converges in one direction, opens in a perpendicular plane, and leaves in films shaped as fans. Droplets form by the impact of the thin liquid film on the surrounding air. Droplet size is determined by the surface tension of the liquid and the pressure at which liquid, suspension, or emulsion is forced through the nozzle. Pressures vary from 25 to 300 p.s.i. and the spectrum of droplets between 50 to 500 $\mu$ diameter (COURSHEE, 1967).

Blast nozzles work by the interaction of air and liquid film, with the air as the moving phase. The liquid film is ejected into the air stream by pressure of one atmosphere or less. The air blast breaks the film into droplets and the droplets are blown toward the target by the sprayer. When blast nozzles are mounted to airplanes, the air blast is caused by the nozzle moving through the air. Droplet size, which depends upon the surface tension of spray and the speed of air blast, falls between 30 and 120 $\mu$ diameter (COURSHEE, 1967).

The spinning disc nozzle works by internal disruption of liquid. This is accomplished by centrifugal atomization (COURSHEE, 1967). The liquid is fed to the surface of a rotating disc or cylinder in which the outermost portions spin off from the body of liquid as drops. Drop size is determined by the surface tension of liquid, speed of rotation and diameter of disc or cylinder. When used in spraying with airplanes, spinning disc nozzles form drops both by internal disruption of the liquid in the nozzle and by the movement of liquid in air. The latter movement is determined, in part, by the air speed of the nozzle.

JOHNSON, KROG, and POLAND (1963) list 39 fungicides that are prepared for application to foliage. Approximately one-third of these are commonly used in the field. Among these are Bordeaux mixture and the fixed coppers, several organomercuries, elemental sulfur, zineb, maneb, ferbam, thiram, captan, folpet, dodine and dyrene. Several oxathiins, Benlate, and 5-*n*-butyl-2-dimethylamino-4-hydroxy-6-methyl pyrimidine, which display systemic action, show promise of becoming effective agricultural fungicides (VON SCHMELING and KULKA, 1966; DELP and KLÖPPING, 1968; and SHEPARD, 1968, respectively).

### Industrial Fungicides

The prominent role that fungi play in biodeterioration necessitates the incorporation of fungicides in coating fabric, wood, leather, textiles, and plastic products. According to WESSEL and BEJUKI (1959), over 400 antimicrobial preparations of 140 chemical combinations are marketed by 171 companies for these purposes. The fungicide varies with the manufacturing process and the use of the product. A recent review by BLOCK (1967) examines the fungicides employed in textiles, paper, rubber, plastics, paints, electrical equipment, petroleum products, leather, and drug and cosmetic products.

Since the methods of applying these fungicides vary with the product and its manufacturing processes, details of application are beyond the scope of this book. Fungicides are incorporated into the product by a processing step where costs of application are minimized and the efficacy of the toxicant is not impaired. Application during a wash, dip, or other liquid process is common.

The treatment of wood with fungicide may vary with the intended use of the timber. BAECHLER (1967) divides the methods into two classes; pressure and nonpressure. In the pressure method, the liquid preservative is injected into wood that is enclosed in a vessel under hydrostatic pressure. The nonpressure method involves dipping or soaking wood in solutions of preservatives or applying the preservative to the surfaces of wood with a brush or sprayer. BAECHLER (1967) lists a third method in which the solution is forced into cavities in wood by the partial vacuum induced in the internal spaces through heating the wood.

A combination of vacuum and pressure methods provides maximum retention of the fungicide, a highly desirable process for preparing marine pilings. High retention in fence posts can be achieved when the posts can take up solutions of fungicide over several months when placed upright in vessels containing fungicide. The treatment is accomplished in shorter time if the fungicide is forced by gravity to flow down the post by maintaining a reservoir of fungicide solution in a cup secured to the top of an upright post (HICOCK, OLSON, and CALLWARD, 1949). Low retention is the order for brush application and short dips. However, fungicide preparations in penetrating oils show high retentive values when applied to the wood surface.

Common fungicides employed in wood preservation are salts of heavy metals, copper naphthenate, pentachlorophenol, and creosote.

## Factors of Dispersal

For discussion, dispersal is restricted to the movement of fungicide from the applicator to the surface to be treated — the diffusion of vapors in soil, the diffusion of liquids in porous substances, and the flight of particles in air. Uniform dispersal is essential to control fungous growth in soil and on propagating material, plants and commodities.

### Diffusion of Vapors

Dispersal and escape of a fumigant is dependent upon its diffusion in air spaces from point of release to distant points in the chamber or soil. The volume of air space is determined by the size and shape of the units of the commodity and the tightness in which they are packed together. Additional space is found in the pores of the units. With soil, the content of free water, which occupies a portion of free space, also affects the amount of air space available for gases to diffuse (HEMWALL, 1960). ECKERT (1967) emphasized the avoidance of overpacking fumigation chambers with fresh produce to ensure air space volume for uniform distribution of gas. HEMWALL (1960) related that the distance a fumigant travels in soil is determined, in part, by its porosity and water content. Hence, the distance between injection, depth of injection and time before planting is dependent upon texture, state of compaction, temperature and water content of the soil.

The amount of fungicide in the gaseous state is affected, in part, by the applied dosage and by the vapor pressure of the toxicant. With sulfur dioxide and chloroamine — toxicants applied in the gaseous state — the dosage is the dominant limiting factor; with the halogenated hydrocarbons, chloropicrin, and methyl isothiocyanate — toxicants applied in the liquid state — vapor pressure limits fumigant action (MOJE, 1960). Low temperature decreases vapor pressure of chloropicrin below that required for dispersal of the fumigant in soil and restricts use of the toxicant to summer treatments (STARK, 1948).

Some gaseous toxicants may arise as products of reactions of the chemical employed. Methyl isothiocyanate is the gaseous toxicant of Vapam and Mylone when these dithiocarbamate materials are applied to soil (HUGHES, 1960, and TORGESON, YODER and JOHNSON, 1957, respectively). The conversion is temperature and pH dependent. Esterification of Vapam prevents the conversion (MILLER and LUKENS, 1966).

The amount of vapor toxicant arriving at the distant points depends upon diffusion of vapor through the substance and adsorption of the gas by the substance. Diffusion rates increase with increase in temperature and decrease with increase in the square root of the molecular weight of the gas. Hence, ammonia diffuses 2.1 times as fast as 2-aminobutane (ECKERT, 1967).

The toxicant is bound by the commodity or soil as the gas diffuses through the interstitial spaces. The bonding is very pronounced, with high ratios of exposed surface area to gas volume. The bonding can be physical, such as dissolving in water or aqueous films that may cover produce and soil particles and in lipophilic coverings of produce. Polar gases tend to dissolve in water and nonpolar gases in lipids. The bonding may be chemical, such as sulfur dioxide forming sulfurous acid in moisture and then combining with cations to form stable salts (ECKERT, 1967). Reactive gases can break down in water of form stable nontoxic products. Reactive halogenated compounds hydrolyze in water (MOJE, 1960). Methyl isothiocyanate forms stable products with thiols (MILLER and LUKENS, 1966). Other sources of lost toxicant are adsorption by packaging material and chamber walls and escape from the surface of soil.

## Diffusion of Liquids

The movement of liquids and solutes to distant points to prevent fungous growth is important in the preservation of wood, leather, and textiles and in the systemic action of fungicides in higher plants. The penetration of wood by liquid formulations of fungicides is capillary and by application of pressure or vacuum (BAECHLER, 1967). Mercury is worked into leather by lipophilic partitioning into the proteinaceous material through surface application of phenylmercury salts of stearic or oleic acids

(ABRAMS, 1948). Copper naphthenate as a solute in organic solvents penetrates cotton fabrics where it is deposited on evaporation of the solvents (BLOCK, 1967).

Systemic action in plants is limited by the entry into the plant and movement of fungicide to sites of fungous attack (CROWDY, 1959). The fungicides enter plants largely under aqueous conditions. Roots, the organs of uptake of fungicides that are applied to soil, grow and function in an aqueous environment. Uptake of compounds by leaves is restricted to wet surfaces. The use of humectants to delay drying of spray droplets enhances systemic action of fungicides (RICH, 1956). Entry into leaves proceeds via stomata or by direct penetration through the cuticle (VAN OVERBEEK, 1956). Stomata are penetrated by vapor, low viscous oils, and solutes in solutions of low surface tension (FOGG, 1948; RICE and ROHRBAUGH, 1953). Cuticular penetration is the common route of entry for leaf and root and the process involves passage through selectively permeable barriers of waxes and cell membranes (CROWDY, 1959).

Translocation of fungicides in plants is through either xylem or phloem, while cell to cell diffusion may account for local systemic action. With xylem transport, movement of compounds is predominantly upward and outward from the stem in the mass flow of the transpirational stream or flow from root pressure (CROWDY, 1959). Tip and marginal burning of leaves following application of fungicide to roots suggests accumulation of toxicants at the ends of the xylem in the leaves. This type of transport is not very useful in distributing the fungicide in the plant because of the rapid accumulation at the extremities of the xylem. To control xylem-invading fungi, however, the movement of cationic fungicides can be slowed down by adjusting the chemical structure of the compound so that the toxicant can bind loosely to anionic sites as the solute moves in the mass flow (EDGINGTON and DIMOND, 1964). The bonding and ease of displacement through cation exchange of alkyl quaternary ammonium compounds is a function of the length of the alkyl chain. Long chains such as dodecyl bind tightly and the compound fails to move up the stem. Movement can be induced by adding divalent cationic metals to the fungicide formulation (EDGINGTON, 1966). Short alkyl chains such as octyl bind weakly and the compound moves to the shoot.

Movement of materials in the phloem involves living cells. Hence, phloem transport stops at sites of injury. Cessation of transport at the site of injury may prove useful for directing fungicides to phloem-invading fungi. Many growth regulators, herbicides, and certain systemically acting fungicides are distributed in the phloem (MITCHELL, 1963). Movement is mostly inward from leaves and downward from stem to roots. Although the mechanism for distributing solutes is partly diffusion, transport is mostly with the mass flow of sugars (MITCHELL, 1963). Thus, the rate of translocation would be greater from mature leaves and in sunlight than from rapidly expanding leaves and in shade (MITCHELL and BROWN, 1946).

Movement from cell to cell in local systemic action is subject to diffusion and permeability of cell membranes of the plant. Movement of fungicides several cells distant may be important for distributing the toxicants from the vascular systems to adjacent tissue. In addition, local systemic action is desirable for foliar eradicants to burn out infections and for protectants to extend the sphere of influence of particles on foliage.

## Flight of Particles

Dust particles and spray droplets are dispersed toward the target, in part, by controlled flight and, in part, by drift flight. The controlled flight is the distance particles travel by momentum when forced from the applicator. Controlled flight of particles from dusters, mist sprayers and smokers is the distance particles are carried in the air jet emitted from the applicator. Controlled flight of particles from pressure sprayers is the distance the water jet and droplets travel after being ejected from a nozzle.

Particles in controlled flight are directed to the target. The success of placing the material on target is influenced by wind and relative humidity. High wind speed can shorten the distance of controlled flight. Gustiness eliminates controlled flight except for coarse sprays. However, wind drift can be used to extend the flight of particles.

Droplets lose water in flight from evaporation and the loss in volume sharply reduces droplet size of those below 100 μ (COURSHEE, 1967). Water loss is dependent upon relative humidity of the air. To reduce water loss by droplets from mist sprayers, low volatile solvents and sugar molasses have been added to the spray (FULTON, 1965). The flight of small droplets is shorter than that of large droplets.

Droplets are projected short distances in air (500 μ droplets, about a meter and smaller ones, a few centimeters) (COURSHEE and IRESON, 1961). Greater effort to extend the projection causes the droplets to break up into smaller ones, which lose momentum quickly. The coverage of a single nozzle is below 2 meters and nozzles are placed on a boom less than a meter apart.

The distance an air jet carries particles is quite variable. The stream is distorted by wind speed and direction and the ground speed of the sprayer (COURSHEE, 1967). Moreover, turbulences within the air jet and at the edge shorten the range of travel.

The distance a liquid jet travels depends upon the angle of the spray jet, output and speed of the sprayer. COURSHEE (1967) expressed it as follows:

$$R = \frac{KV^{0.5}U}{\theta S}$$

where $R$ = range (meters); $\Theta$ = angle of spray jet (degrees); $V$ = spray output (liters per second) at $U$ speed (meters per second); $S$ = sprayer speed (meters per second). In short, a wide angle nozzle projects short distances for directing spray to ground crops and an acute angle projects long distances for directing spray to tree tops.

In controlled flight spraying, the material is applied directly to crops with little chance for spray to drift away from crops. Especially with coarse nozzles, the operation can be accomplished in spite of considerable wind. However, the width of swath through a field is limited to the length of the spray boom. Moreover, with pressure spraying, large volumes of water are employed and the equipment is elaborate.

Once momentum of controlled flight is lost, particles drift in the air. Controlled flight may direct particles to the outer leaves of plants, but it is drift flight that carries them to inner leaves (FULTON, 1965). With dusters and mist sprayers, particles directed toward the target actually reach the surfaces of plants by drift flight. Because of eddy effects, only a portion of drifting particles are entrapped in the quiet air next to leaf surfaces, the rest passing on to distant surfaces. Hence, deposition on all the leaves of a plant approaches uniformity.

Drift application of fungicides has many advantages. Both large and small particles drift at the same rate, thus giving a unified deposition of particle size across the swath of the sprayer. This is not so with particles in direct flight, where most large particles are found in the middle and small particles at the edges of the swath. With drift spraying, droplets diffuse at an average angle of 7° to either side of vertical. The divergence gives a wide spray swath to each pass of the machine (COURSHEE, 1967). Moreover, the formula is concentrated in small volumes of water, thus permitting the use of light-weight equipment. Misting machines can be worn as knapsacks for treating small plots or can be mounted on aircraft for treating extensive areas. The disadvantage of drift application is that spraying or dusting is limited to times of quiet air movement. For discussion of low gallonage and drift spraying see COURSHEE (1967) and FULTON (1965).

## Factors of Deposition

The amount of fungicide deposited and its coverage on the surface are highly influenced by the surface itself and the form of the fungicide which hits the surface.

### The Treated Surface

The heterogenic composition of surfaces of plants, fabrics, and soil particles may cause difficulty in coating of the surface with fungicides (DIMOND, 1962). Leaf surfaces of apple, bean and tomato are hairy, while those of banana, celery, pea, and crucifers are quite smooth. Leaves of many grasses contain high ridges. All leaves are porous because of their stomata. All leaves harbor dust or foreign particles which affect the deposition of fungicides. Hairs, ridges, pores, and dust produce a rough surface.

Although surface roughness may aid deposition of fungicidal dusts, it hinders deposition of aqueous preparations. Roughness counters wettability because air entrapped in the cavities reduces the contact area between the surface and the liquid. A smooth surface may be perfectly wettable to water, but a rough surface of the same composition makes a contact angle with water above 100° and repels the water. The degree of water repellence, other things being equal, is related to the uniformity and persistence of the roughness, providing that the dimensions of cavities of roughness are great enough to avoid seepage of water from capillary action (ADAM, 1958). RICH (1954) found a higher contact angle between water and surface and less deposit of zineb on hairy leaves of bean than on smooth leaves of celery following sprays of the fungicide.

Surfaces of leaves contain a layer of epidermal cells and a cuticle. The walls of the epidermal cells contain cellulose and openings exposing the protoplast. The exposed protoplasts as well as stomata may serve as sites of entry into the plant (VAN OVERBEEK, 1956). The cuticle contains platelets of waxes embedded in and rodlets of waxes protruding from a matrix of cutin and pectins. Cellulose, cutin, and pectin affect hydrophilic properties of the surface. The wetting properties of leaves have been associated with the occurrence of these waxy projections (JUNIPER, 1959). The negative charges of cell wall material may serve as electrokinetic binding sites for positively charged fungicides (RICH, 1954).

The composition of leaf surfaces changes with age, nutrition and weather during development. Hairs and cuticle are less developed on young leaves than on mature leaves. The cuticle is thinner on young and succulent leaves than on mature and hardy leaves and, thus, has greater hydrophilic properties. Leaves formed during cloudy weather may have less wax and smoother surfaces than those formed during sunny weather. Surfaces with little wax and smoothness are more easily wet than those with a lot of wax and roughness.

With aqueous sprays, the shape and orientation of plant leaf determines, in part, the volume of water film a leaf surface can hold before runoff occurs. Most leaves are planar and have edges. The leaf edge can serve as a dam to hold back the water film. A horizontal edge is a better dam than is a vertical edge and a continuous edge is better than a discontinuous one. With a discontinuous edge, points of leaf edges serve as channels for water flow over the dams. A conical shaped leaf with no edges would have little volume in its water film prior to runoff. An onion leaf, besides having a smooth waxy surface, is conical and vertical and thus is extremely difficult to cover protectively with a fungicide.

## The Fungicide Form

Although vapors kill fungi quickly, their action may be delayed by adsorption of toxicant on particles of the medium. The adsorption is dependent upon the total surface area of particles. Surface roughness and porosity increase adsorption to dry surfaces, while solutes, pH, and surface tension of water films affect adsorption to moist particles.

Solutions of fungicides wet the treated surface and deposits of fungicides adhere to surfaces of fibers and other minute constituents of the surface on evaporation of the solvent. Low surface tension of solvent and smooth surface hastens the wetting process. Forces of adherence of fungicide to surface may be of an electrolytic, electrokinetic, or a nonpolar nature (RICH, 1954).

The deposition of fungicides in particulate form is complicated by the nature of the particle, by contact of particle with the surface, and by the fate of the particle following contact.

According to COURSHEE (1967), particles land on a surface by one of four processes: sedimentation, attraction, interception, or impaction. Particles settle on horizontal targets at terminal velocity. Small particles carried in moving air tend to travel with air as it moves around the target. Extremely small particles (below 30 μ diameter) with static charges may be attracted to the target as the air mass moves over the target. Large particles, because of their momentum, may continue on a straight line course to hit the target in spite of change in direction of air flow around the target. Direct impact is about 50% for particles over 30 μ diameter moving in air (COURSHEE, 1967).

With dusts, the particles are electrostatically charged on leaving the duster and during forced flight through the air. The nature and influence of charge on the particle depends upon particle size, speed of duster, atmospheric conditions, diluent and impurities in the formulation (WILSON, JANES, and CAMPAU, 1944). Poorly charged particles suspended in air aggregate and fall out of the air. Highly charged particles remain dispersed for prolonged drift flight, which is necessary for thorough coverage of plant surfaces.

Droplets of water pick up negative charges from the water surface as they are sheared from that surface (LOEB, 1958). The droplets pick up additional static charges when propelled through the air. The charge on droplets of less than 30 μ diameter may be sufficient for attracting water droplets to target from the air.

The angle of contact between the droplet and leaf surface determines the area wet by a spray droplet. On smooth waxy leaf of pea, BRUNSKILL (1956) found that droplets bounced off after impact when the contact angle was high and droplets remained intact. No wetting occurred with bouncing. Bouncing increased with droplet size and surface tension of the solvent, and as the angle of incidence approached the perpendicular. All droplets of 243 μ diameter and surface tension of water (72 dynes per cm.) bounced. However, only 70% of droplets of the same size bounced off the leaf when the surface tension was reduced to 45 dynes per cm. Obviously, droplet bouncing curtails deposition of fungicide. Small droplets that fail to bounce but retain a spherical shape may run off inclined surfaces to reduce deposition (HORSFALL, 1956).

At the other extreme of acute contact angles, droplets spread out into a film. The films of many droplets coalesce to cause a flow of solvent off the leaf. Part of the fungicide is lost in the drain-off. Thus, as the contact angle is reduced, droplet spread increases, and retention passes through a maximum range and declines (RICH, 1954). The use of surfactants to achieve proper spreading of droplets can lead to drain-off if the nature of the plant surface is ignored.

Spray droplets may coalesce in spite of proper spreading if their deposition becomes congested. Congestion occurs from over-spraying or uneven spraying. Outer leaves may be over-sprayed in order to cover completely the inner foliage. Uneven spraying results from an uneven pattern of spray droplets emitted into the air.

The ability of the spray droplet to remain flat or in a film on a surface is known as its wetting property (MARTIN, 1964). The wetting property is measured by the contact angle between liquid and solid at the receding front. The receding concact angle is always less than that of the advancing front. Numerous reasons have been proposed to explain the apparent increase in hydrophilicity of the wetted surface of solid. CASSIE (1948) suggested that water replaces air in the crevices and cavities as the droplet spreads over the surface. When the film recedes, the effective surface of the solid has changed to one more hydrophilic. Hydrophilicity may be increased by the adherence of spray surfactants to plant surface with the polar ends outward (EVANS and MARTIN, 1935) or by the overturning of hydrophobic materials of the plant surface when in contact with water so their polar ends point outward (LANGMUIR, 1938). ADAM (1964) suggested that liquids may remove impurities from the surface of the solid.

A complete coverage of plant surface is not necessarily required for disease control (COURSHEE, 1967; FULTON, 1965). Deposition of air-borne pathogens is not at random. GREGORY (1961) found that the deposition of mildew conidia to surfaces from air was affected by shape of surface, its position in respect to air currents, and movement of air. Patterns of spore deposition are similar to those of spray droplets from drift flight. HARTILL (1968), by choice of nozzle, position of nozzle, and spray pressure, was able to match deposition of spray droplets to that of air-borne spores of *Eryshiphe cichoracearum*, *Alternaria longipes* and other fungi on tobacco leaves. Matching spray deposition to spore deposition minimizes coverage and the amount of deposit required to control disease. By restricting spray deposition to sites of infection on the plant, the means of chemical control of banana diseases was altered to less expensive operations (FULTON, 1965).

Water-borne spores or spores requiring free moisture for germination may attack the tips of leaves, the site where dew drops and water from other sources accumulate. During epidemics, conidia of *Phytophthora infestans* germinate as zoospores in drops of water at the tips of potato leaves. Epidemics of late blight disease can be prevented if coverage is restricted to the tips of leaves (COURSHEE, 1967). The crude application of Bordeaux mixture with a watering can suffices for depositing fungicide at the tips of leaves as well as the more sophisticated application with a mist sprayer. However, the former technique is wasteful because most of the fungicide applied to the leaves drains off to the ground.

## Factors of Retention

Retention of fungicide designates the amount of fungicide remaining on a surface following application. MARTIN (1964) has called this initial retention, thus distinguishing the initial deposit from that persisting with time. The nature of the fungicide and the nature of plant leaf affect the amount of fungicide retained.

### The Nature of the Fungicide

Retention of copper by bean leaves is maximum when the metal is part of a hydrogel residue (EVANS *et al.*, 1962; BURCHFIELD and GOENAGA, 1957). Bordeaux mixture forms the gelatinous residue when low amounts of lime are used (10:3 copper sulfate to lime by weight) or when freshly prepared ratios of high lime are used. Stale preparations of the 10:10 ratio grow large sphaerocrystals at the expense of the hydrogel and deposition falls off (BURCHFIELD and GOENAGA, 1957). Crystallite growth is retarded by inorganic ions, sugars, amino acids and ligno sulfonates (BURCHFIELD, SCHECHTMAN and MAGDOFF, 1957). Copper oxychloride, which remains largely in particulate form in residues when the copper salt is mixed with lime and clays, causes less deposition of copper on foliage than does Bordeaux mixture (EVANS *et al.*, 1962). However, copper oxychloride forms gelatinous residues with aluminum hydroxide and the retention of copper from this mixture is similar to that of Bordeaux mixture.

Apparently, the strong cohesive bonds of the colloid which holds the deposit together and the adherence of the deposit to the leaf surface make possible the stable buildup of Bordeaux mixture on the plant (RICH, 1954). RICH suggested that Bordeaux mixture adheres to leaf surfaces by electrokinetic bonds and the number of electrokinetic anionic sites on the leaf delineates retention. However, the electrokinetic theory is weakened by the retention pattern of copper oxychloride to leaves (EVANS *et al.*, 1962, 1966). Particles of copper oxychloride are positively charged, but only when in a hydrogel condition do they build up deposits as Bordeaux mixture does. Apparently, what limits the maximum deposit of Bordeaux mixture is the thickness of the hydrogel on the leaf surface.

Wetting agents can increase, decrease, or not affect initial deposition of the fungicide (Somers, 1957; Burchfield and Goenaga, 1957; Evans *et al.*, 1966). The action may depend upon whether the wetting agent improves spread and retention or causes excessive drain-off of spray.

The concentration of fungicide in the spray tank and the total gallonage applied influence the retention of fungicides on leaves. The retention of zineb and copper oxychloride increases with the concentration of the fungicide in the spray tank (Rich, 1954; Evans *et al.*, 1962), while the retention of Bordeaux mixture can increase with the logarithm of the concentration of fungicide in the spray tank (Rich, 1954). Retention of copper falls off as the volume of water applied per land area increases (Evans *et al.*, 1966). Apparently, retention of copper is limited by the drain-off problem. Thus, small spray droplets and drift spraying ensure retention of high amounts of fungicide by plant foliage.

### The Nature of the Plant Surface

Many experiments have shown that initial retention can be affected by the type of leaf surface (Rich, 1954; Somers, 1957; Evans *et al.*, 1962, 1966). Failure of droplet spread on hairy leaves of bean may have caused low initial retention of zineb in relation to the retention of the fungicide on celery leaves (Rich, 1954). When surface tension of spray liquid is high, retention increases with decrease in that characteristic of the spray liquid (Furmidge, 1962). However, high surface tension may curtail retention on smooth waxy leaves by causing the spray droplets to bounce off leaves instead of spreading (Brunskill, 1956; Burchfield and Goenaga, 1957; Evans *et al.*, 1962). Apparently, a high surface tension has little effect on the spread of Bordeaux mixture on rough and smooth leaves (Rich, 1954), but, under certain conditions of application, drops of Bordeaux mixture bounce off smooth waxy leaves (Burchfield and Goenaga, 1957; Evans *et al.*, 1962).

## Factors of Persistence

Persistence describes the amount of fungicide remaining in a protective barrier over a period of time. Elements of weather and expanding leaf surfaces tend to deteriorate the barrier. Resistance to physical removal of the barrier (tenacity) and the resistance to chemical deterioration of toxicant in the barrier are characteristics of the fungicide that affect persistence.

### Resistance to Physical Removal

A tenacious fungicide can persist in a protective barrier through all types of weather. Good tenacity is highly influenced by the physical state of the residue. The excellent tenacity of Bordeaux mixture lies in the strong cohesive forces between the colloidal particles which resist the pulling apart of the colloid when adsorbed to the leaf surface or the dissolution of the dried gel (Rich, 1954).

The rate of leaching of copper from residues of Bordeaux mixture by rain is less dependent on the amount deposited than it is on weather conditions. Rich (1954) found that percentage loss from weathering decreased as the amount deposited increased, while BURCHFIELD and GOENAGA (1957) found that the logarithm of tenacity was related to the intensity of rain. This type of loss is characteristic of homogenous spray deposits.

Loss of fungicide from particulate residues is exponential with deposit (RICH, 1954; BURCHFIELD and GOENAGA, 1957). That is, most of the fungicide is lost quickly while a small remnant is held fast to the leaf. The type of loss suggests that cohesion between particles is weak and adhesion of particles to the leaf is strong. Moreover, the loss attests to variation in adherence of the components of the residue to the leaf.

All of the fungicide dislodged by rain is not entirely lost by the plant. About 10% is redistributed to unprotected areas (HAMILTON, MACK, and PALMITER, 1943). Sulfur, 40 $\mu g$ per $cm^2$ can be redistributed by rains following application. Ferbam and dodine are redistributed as effectively as sulfur, while captan and glyodin are poorly redistributed (HAMILTON, 1968). Redistributed materials can be transferred over the leaf surface from upper to lower surface and from one leaf to another.

Plant surfaces expand rapidly during bud opening and leaf development. An expanding leaf surface creates unprotected areas in the barrier of fungicide. Moreover, an expanding surface can dislodge fungitoxic residues that were present. New leaf area can be protected in part by the redistribution of fungicide in rain. However, to maintain a protective barrier during the early growing season when buds are opening and surfaces are expanding, frequent application (every 5 to 7 days) is required. Later, when leaves and fruit have completed their growth, an occasional spray of fungicide will suffice to maintain a protective covering on plant surfaces.

## Resistance to Chemical Deterioration

Many potentially effective fungicides have failed in the field in spite of laboratory tests attesting to the chemical stability of the compounds. The discrepancy may lie in the difference in form of the chemical in the two cases. In the residue where the chemical is distributed in a thin film or minute particles, the functional surface area of the chemical is extremely high. Functional surface area of the chemical can be low when particles are packed together or suspended in flakes, which may be the case in laboratory tests. Thus, reactions that deteriorate the fungicide may proceed rapidly in field residues and evade detection in laboratory tests. Stability tests should mimic the state of the fungicide in surface residues.

The toxicant can deteriorate in the barrier by vaporization, photolysis, and hydrolysis. Although vaporization from solids is influenced by surface area, weight loss of very small particles is actually proportional to their radii (BURCHFIELD, 1967). Usually, with compounds of low vapor pressure, the diffusion of vapor away from the air adjacent to the solid limits evaporation. The action of particulate fungicides has been shown to proceed in part, by vapor (THATCHER and STREETER, 1925; MILLER and STODDARD, 1957). Injury to turf by phenyl mercury fungicides at temperatures above 30 °C (WADSWORTH, 1960) suggests a switch to vapor activity with increase in temperature. The restriction of effectiveness of chloranil to cool season crops and dichlone to spring application suggests that both toxicants may be lost from sublimation. This is consistent with the sublimation temperatures of 24 and 29 to 32 °C, respectively, for chloranil and dichlone (RICH, 1968).

Colored compounds absorbing light in the visible and near ultraviolet regions are susceptible to photo decay. Yellow pigmented substituted quinones and Dexon in residues have been reported to break down in sunlight (BURCHFIELD and MCNEW, 1950, and HILLS and LEACH, 1962, respectively). In aqueous solution, quinones are photo-converted to hydroquinones and dimers when exposed to wavelengths below 577 mμ. Chlorination of quinone retards decay by light and the photo-action spectrum shifts toward longer wavelength with increase in oxidation potential of the quinone. Thus, photo-stability decreases in the order: dichlone > chloranil > quinone. With Dexon, photo-decomposition proceeds by two stages: (1) reduction of the diazonium group with evolution of nitrogen and then (2) oxidation and polymerization of the products of the first reaction. Fungitoxicity of chloranil and Dexon is destroyed in sunlight.

Fungicides containing ester or ether links or functional groups of amide, nitro or halogen can hydrolize in water to less potent materials. Fungicides reportedly hydrolyzing from residues are captan, folpet and, possibly, dichlone. The basidity and surface tension of the slurry, choice of diluent in spray formulation, and cationic contaminants of the spray water affect hydrolysis of captan in spray residues (DAINES *et al.*, 1957). Apparently, folpet in residues is more susceptible to hydrolysis than captan because folpet in solution hydrolyzes faster than does captan (LUKENS and HORSFALL, 1967). Dyrene, which contains reactive chlorine atoms, hydrolyzes slowly in water (one half-life = 20 days compared with 2.5 h for captan) (BURCHFIELD and SCHECHTMAN, 1958) and hydrolysis of the compound does not appear significant in residues.

## Factors of Formulation

Formulation is concerned with presenting the fungicide in the most effective form for storage, application, performance, and least cost. Fungicides have been prepared as solutions, emulsions, wettable powders, and granules. Wettable powders are the most popular form used in plant protection.

### Particle Size of Water Insoluble Formulations

The ability of a fungicide to protect the surface from fungous attacks depends more upon the number of fungicide particles per unit area than upon weight per unit area (WILCOXON and MCCALLAN, 1931). Thus, efficacy increases with decrease in diameter of the fungicide particle. The reduction in particle size aids dilution and improves coverage of the fungicide.

However, a minimum particle size is reached beyond which efficacy does not increase and may fall off (BURCHFIELD, 1967). Maximum efficacy is accomplished when all loci on the surface for fungous attack are covered by the influence of particles of fungicides. When these loci are protected by extremely small particles of fungicide, efficacy may be curtailed by destruction of the fungicide. Reactions of decomposition, which depend upon surface area of fungicide exposed to the elements, increase with reduction in radii of the particles (BURCHFIELD, 1967). Dusts and wettable powder fungicides are ground to mean radii in the range 1 to 10 μ (SOMERS, 1967, and BURCHFIELD, 1967).

## Diluent

Since the amount of fungicide required in the residue is minute (5 to 50 μg per $cm^2$), safe use of the toxicant calls for great dilution in application. Water is the major diluent, which the applicator adds to the mixture at the time of spraying. Wettable powder and dust formulations contain some inert clays as diluents. These are selected on the basis of cost, low cation exchange capacity, lack of objectionable impurities and physical characteristics suitable for grinding into powder. Other materials for use as diluents have been several metal carbonates and oxides, micronized sulfur, and gypsum (SOMERS, 1967). Granular formulations may utilize ground corn cobs, vermiculite, particles of plastics and sand as well as clays. The diluent may be used to merely dilute the particles of fungicide or may serve as a carrier with the fungicide adsorbed to the particles of diluent (DUYFJES, 1958). Although diluents are considered inert, they can catalyze hydrolysis of chlorinated fungicides to hydrochloric acid. Hydrochloric acid produced in spray residues can be phytotoxic (DAINES *et al.*, 1957; MILLER, 1957; BURCHFIELD, 1967). Reduced performance may occur when cationic fungicides bind to clays or organic diluents (ARK and WILSON, 1956).

## Surface Active Agents

Wetting agents are incorporated into wettable powders to ease the wetting of these formulations in the spray tank. Wetting agents are incorporated into solution and emulsion spray formulations to provide good penetration of disease lesions, textiles, paper, or other material being treated. However, it is unfortunate that wetting agents employed in wettable powders actually reduce droplet size, shorten flight, and curtail deposition of material from spray droplets on the plant surface (DIMOND, 1962). As the surface tension is decreased, less volume of spray is retained on the plant surface. In addition, the presence of a wetting agent in the dry residue hastens rewetting and attendant erosion and decomposition of the toxic agent. Wetting agents in agricultural sprays are usually small molecules of the nonionic detergent-type surfactants.

Spreading and sticking agents, in contrast to wetting agents, are surface active agents of large molecular size, such as proteinaceous materials, polymeric lattices or highly viscous oils (SOMERS, 1967). The compounds aid spread of spray droplets through lowering the surface tension of the spray liquid. They aid tenacity of the spray deposit by forming a cohesive gel over the dried particular deposit (MARTIN, 1964). Spreading and sticking agents resist rewetting during short wet periods because they are not very soluble in cold water.

## Safeners and Adjuvants

Various materials have been incorporated into formulations to stabilize the active ingredient in storage, to prevent decomposition in residue or to neutralize obnoxious decomposition products. Urea has been used in trace amounts to avoid autolytic breakdown of halogenated toxicants in storage. Calcium carbonate, magnesium oxide, and other basic salts neutralize hydrochloric acid produced in wet residues of captan (DAINES *et al.*, 1957; MILLER, 1957). Golf course superintendents have a practice of adding thiram to phenyl mercury sprays to avoid injury from mercury during hot weather (WADSWORTH, 1960). Glycerol and other humectants, when added to aqueous sprays of chemotherapeutants and herbicides, enhance uptake of the active ingredient by the plant, presumably through retarding the drying of the spray residue (GRAY, 1956; and RICH, 1956).

## Conclusions

Fungicides are applied to surfaces to prevent attacks by fungi. The performance of the barrier depends upon the area influenced by minute deposits of fungicide. Potency, coverage, and persistence of the deposited fungicide determine the effective life of the protective barrier.

The fungicide is dispersed in a vapor, liquid, or particulate form to the treated surface. Numerous methods and machines have been devised for application. The method and conditions of application affect coverage of fungicide over the surface.

The fungicide and its formulation have a profound influence on the coverage and persistence of the protective barrier. The surface tension of spray liquid influences the size of droplet, its flight pattern and degree of bouncing or spreading over the treated surface, and the amount of fungicide retained by the surface. Materials that form hydrogels or other matrices enable the barrier to withstand loss of effectiveness, redistribution and decomposition from weathering, and crumbling from enlargement of the treated surface. Diluents help attenuate the active ingredient or its obnoxious decomposition products. Humectants delay the drying of residues after application, thus enhancing the uptake of systemically acting fungicides by the plant surface. A small particle makes possible effective coverage with small dosages of active ingredient.

The treated surface plays an important role in coverage and persistence of the protective barrier. Size, shape, and orientation of surface affect the interception of air-borne particles. Composition and degree of roughness affect bounce of particle and spread of particle over the surface. The chemical composition of the surface determines, in part, the nature of adhesive bonds between the barrier and the treated surface. The flexibility and expansion of plant surface affect the persistence of the barrier on the plants.

Chapter 4

# Migration of Fungicide to Sites of Action

## Introduction

Fungicides migrate to sites of action where reactions with cellular constituents inhibit further growth of the fungus. With protectant fungicides the entire process has to occur on the host before infection occurs. The migration of fungicides proceeds in spite of the ability of many fungous spores to survive in a hostile environment. During migration the fungicide in residues is mobilized and permeates to sites of action on or in the fungous cell. The mechanisms involved will be discussed in view of recent concepts of chemical activity of fungicides and of cellular organization of fungi.

## How the Toxicant Becomes Available

How do fungicides come in contact with fungous cells? This phenomenon has interested plant pathologists since the early use of foliar sprays. Though physical contact between particles of fungicides and fungous cells often occurs, little fungicide in the solid state is absorbed by the fungus. The fungicide in the residue must be transformed from a solid to an active state in order for it to be useful. Fungi contact fungicides in the vapor state or as solutes.

### Vaporization

Fungi contact volatile materials in the gaseous state. These are the soil and chamber fumigants — chloropicrin, methyl bromide, dichloropropane, and sulfur dioxide. Pentachlorphenol, naphthalene, and sulfur are absorbed by fungi in the vapor state. The selection action of organo-mercurials is distinguished by the vapor pressure of the mercurial.

However, contact by vapor has been little appreciated for organic fungicides which have seemingly low vapor pressure. Dichlone, which sublimes above 30 °C (Rich, 1968), may be contacted by fungous spores on plant leaves as a vapor from residues of dusts and sprays. Thiram, in dry deposits and with no visible contact with fungi, inhibits germination of spores in the absence of free water (Brook, 1957).

The availability of vapor to fungus is determined in part by (1) the rate of vapor production, (2) the rate of transport of vapor through space in the medium and (3) the competition between fungous cells and other components in the system for the vapor.

The rate of vapor production is governed by the amount and distribution of toxicant in the environment, the vapor pressure of the toxicant, and temperature.

Distribution of fungicide and factors affecting diffusion of vapor have been discussed in Chapter 3.

A rapid diffusion of vapor through the medium is desirable to reach an effective concentration for fungous control. However, unless the volume is contained, as in gas chambers, the rapid rate of transport can be a means of escape for the toxicant. Because of escape, the effective zone of pest control in soil is actually reduced as soil air space is increased (HEMWALL, 1960). Since most soil pathogens flourish in an aqueous environment, soil moisture is important in vapor transport and contact. An increase in Henry's Law constant (the ratio of fumigant concentration in water to that in air) would tend to retain a reservoir of toxicant in the water phase. Vapors of high molecular weight toxicants are retained in water longer than those of low molecular weight toxicants because diffusion coefficients of the vapors decrease with increase in molecular weight. Moreover, diffusion coefficient is reduced further by the presence of a companion vapor of molecular weight higher than that of air. A volatile companion of high molecular weight may increase efficacy of volatile toxicants on plant foliage. Apparently, the reduction in efficacy by an increase in diffusion coefficient from heat is offset by an increase in vaporization of toxicant by heat. The efficacy of sulfur increases with temperature (HORSFALL, 1956). However, an increase in temperature may increase fungous growth and thereby decrease efficacy of a fumigant (NELSON and RICHARDSON, 1967). The higher the proportion of available vapor that is absorbed by the fungus, the more effective is the inhibition. However, the escape of vapor from the system and absorption by other components reduces the amount available to the fungus. The toxicant escapes from the fungus through leaks in chamber, by diffusing in a high volume of air in soil, or by being carried in wind from soil or leaf. Plant material, organic matter and soil particles may decompose or detoxify the toxicant.

## Solution

*In Aqueous Media.* Fungous spores play a direct role in solubilizing fungicides from residues. How copper becomes available from residues of Bordeaux mixture was a puzzle for many years. A water film is necessary for a fungus to obtain copper from Bordeaux because hyphae can grow on dry deposits (BROOK, 1957). However, copper is difficult to leach from residues of Bordeaux. GOLDSWORTHY and GREEN (1936) reported that fungous excretions are required to solubilize the metallic ion. Malic and amino acids were found among the excretions of copper-sensitive fungi (MCCALLAN and WILCOXON, 1936). The toxicity of copper oxides has been related to the chelation of copper from oxides by glycine (HORSFALL, MARSH and MARTIN, 1937). Though host tissue may excrete chelating compounds on interactions with fungi, the major source of copper solubilizing compounds is considered to be the fungus (HORSFALL, 1956).

The new organic protectants, though considered insoluble in water, are soluble to a few parts per million. Those that are not volatile may rely on an aqueous contact for action. Solubilization of these toxicants can be increased by a reduction in particle size in the protective coating. Fungous excretions mobilize thiram by reducing that compound to the dimethyl dithiocarbamate anion (RICHARDSON and THORN, 1961).

*In Nonaqueous Media*. Lipophilic organic fungicides may be contacted by fungi through the cuticle of foliar surfaces. Fungicides may partition from the surface residue into the wax of the cuticle and diffuse in the oil phase to distant parts of the cuticle. The fungus can contact the toxicant during penetration of this layer. The cuticle may serve as a medium of contact for nonvaporous mildewcides (LUKENS and HORSFALL, 1967). The relationship of chemical properties of folpet, captan, and related compounds to the control of powdery mildew of snapbean indicates that the intact molecule is the toxic agent in each case, and not a volatile decomposition toxicant common to all of these compounds. The fact that control of mildew was related to the partitioning of fungicide from water into oil is indicative of the fungicides partitioning into cuticular waxes where the fungus absorbs the fungicide. The mildew spore germinates and penetrates bean tissue in the absence of free water.

Fungicides effective against subcutaneous infections of dermatophytes must penetrate proteinaceous and fatty barriers of animal skin. Hyphae of the pathogen growing in these tissues may contact the fungicide in absence of water.

## How the Toxicant Enters the Cell

In this discussion, permeation will include uptake, penetration of cellular membranes, and migration through the protoplast.

### Removal of Toxicant from the Ambient Medium

During the past 40 years, various workers have shown that fungicides are taken up rapidly by fungous cells from the ambient fluid. This rapid uptake differs from normal diffusion patterns. Studies have shown that the diversion from diffusion can be accounted for by several mechanisms — some physical, some biological, and some a combination of both — that remove fungicide from solution within.

From an extensive series of studies with radioisotopes of metallic ions and certain organic fungicides, MCCALLAN and MILLER (1963) have established that toxic dosages accumulate within a few minutes' exposure of spores of several phytopathogenic fungi. This quick accumulation can reach 10,000 times the concentration in the ambient fluid. The fast accumulation of large amounts of fungicides appears to be a prerequisite for toxicity to spore germination. On a weight basis, conidia of *Neurospora sitophila* take up 13,000 μg/gm (approximately twice the ED-50 dose) of glyodin in one-half minute when the external concentration of fungicide is 34,000 μg/gm of toxicant. Fifteen minutes elapse before the amount absorbed is doubled. Ninety percent of maximum of dodine is accumulated by yeast cells in one minute (BROWN and SISLER, 1960). The absorption of captan, dichlone, ferbam, silver, mercury, and cupric ions is rapid, also. However, cadmium and zinc are taken up slowly and the total amounts are less than those for other fungicides (MCCALLAN and MILLER, 1963).

Numerous reports have shown that the environment plays a direct role in the removal of fungicides from the exterior. Often a variation in applied dosages changes the rate of absorption. The ratio of amount taken up to the applied dosage becomes less with increase in the applied dosage. The pH of the ambient environment affects uptake. Accumulation of dodine by conidia of *Neurospora crassa* increases and then decreases with increase in pH, with the maximum occurring in the range of pH 8 to 11 (SOMERS and PRING, 1966). However, uptake of captan by the same fungus is not affected by the external pH when loss of fungicide from hydrolysis is accounted for (RICHMOND and SOMERS, 1962). The removal of captan, cupric ions, and glyodin from solution by fungous cells increases with temperature, while that of dodine is not thus affected. A sharp temperature dependency implies that absorption is largely by biological and chemical mechanisms, while the reverse implies that absorption is largely by physical means such as adhesion to cell wall or membranes.

Compounds in the medium with binding sites common to those of fungicides interfere with absorption. Cationic compounds, which are removed from solution by cation exchange, can impede the absorption of cationic fungicides. However, the displacement by cationic substances of toxicants from treated spores is not universal. Certain toxicants bind irreversibly with cellular constituents.

Apparently, absorption of fungicides by fungous cells is passive and is governed by the ability of the fungus to accumulate and to utilize the toxicant. A high applied dosage can saturate the fungus with toxicant. The quickness of the uptake indicates that there are no blocks in the path of uptake and cellular sites for attachment and reaction are available in large amounts to the fungicide. The fungicide may be consumed in part by bonding to polymeric structures in the cell wall, membranes, and protoplasmic organelles; by storage in free pools of metabolites and food reserve; and by consumption by metabolic functions of catabolism and detoxication.

From studies of several fungicides and several species of fungi, MCCALLAN, BURCHFIELD and MILLER (1959) have distinguished two types of relationships between uptake and external concentration: uptake is proportional to external dosage; and uptake is proportional to log of external dosage. The arithmetic relationship is descriptive of a situation where the quantity of receptor sites is unlimiting. Toxicants that enter second order reactions may approach an arithmetic relationship if reactions consume the major part of that absorbed and the renewal of the cellular reactive group is not limiting. Uptake is proportional to the log of the external dosage when the number of binding sites are limited, *e.g.*, the adsorption of fungicide to cell walls. The dosage-response curve may vary with fungous species and even with cell type within a species.

## Penetration of External Barriers

Fungicides, except those possessing a cationic charge, readily pass through the walls of fungous cells. The wall consists of loosely knitted fibrils containing chitin and cellulose or other glucans and cementing materials containing protein, polysaccharides (*e.g.*, glucans, mannans, and galactans), and lipids (BARTNICKI-GARCIA, 1968). Apparently, acidic groups, among others, contribute to the electronegative charge that is characteristic of the surface of fungous spores (DOUGLAS, COLLINS, and PARKINSON, 1959).

The binding of fungicides by conidial walls varies with cationic toxicant and with fungous species. Using radioactive tracer detection, OWENS and MILLER (1957) found only a small trace of heavy metals (Table 1) in conidial walls of *Neurospora sitophila*. Their data indicate, however, a 25 to 50% concentration of metals in conidial walls of *Aspergillus niger*. Cell walls comprise 35% of dry weight of both species. SOMERS [1963 (1)] found 25% of a toxic dose of copper in conidial walls of *N. crassa* and 33%

in conidial walls of *Alternaria tenius*. About 50% of a toxic dose of dodine was found in conidial walls of *N. crassa* [Somers, 1963 (2)], while only 3% of a toxic dose of glyodin was found in conidial walls of *N. sitophila* (Owens and Miller, 1957). The wide variation between the binding of these two cationic surfactant toxicants to conidial walls of these two closely related fungous species is surprising. It is unfortunate that both toxicants were not tested on the same test organism under comparable conditions to determine whether the discrepancy lies in test organisms, fungicides, or technique of determination. Neither Somers nor Owens and Miller considered that part of a toxic dose bound to cell walls to be important in the action of the fungicide.

Cell walls of ascospores of *N. tetrasperma* have a high capacity to adsorb substances. This is not due to any metabolic function. Sussman and Lowery (1955) have found uptake of methylene blue to be identical in both dormant and metabolically active ascospores. The nature of elution of the dye with metallic cations suggests that the uptake is by base exchange mechanisms. Also, metals reduced uptake of the dye, presumably by competing with the dye for binding sites. Sussman, von Boventer-Heidenhain and Lowery (1957) have shown that cells of dormant ascospores can adsorb a toxic dose of cationic toxicants. With respiration as a criterion of viability and breaking of dormancy, they found that no toxic action occurs until after dormancy is broken and metabolic activity is initiated. They concluded that the toxicants are retained in the cell walls and do not penetrate the protoplast until dormancy is broken. The authors calculated that the amount of silver bound to ascospore walls would form a layer 7 cations deep if all binding occurred on the spore surface. However, the total active surface of the cell wall is many times the apparent surface area of the spore.

The composition of the cell wall determines the capacity for adsorbing cations by ascospores (Sussman, 1963). Contrasted with conidia, which have one layer of cell wall, ascospores have a complex three-layered wall. The wall is highly lamellar. The outer layer consists largely of uronic acid residues, the assumed binding sites for cations.

## Penetration of Cytoplasmic Membranes

The selective permeability of the plasma membranes of bacteria and cytoplasmic membranes of cells of other organisms is well known. The presence of these membranes has been documented by studies of osmotic phenomena, electron microscopy, and isolation. The nature of the transport of compounds across membranes has been vigorously pursued in recent times. Reviews are abundant. One review, which stresses the nature of membranes in relation to the permeation of fungicides, is called to the reader's attention (Miller, 1962).

Apparently, the function of the cytoplasmic membrane is to allow full passage of desirable materials into the protoplasm and at the same time to exclude undesirable materials. Thus, for passage of toxicants, the materials must be disguised in the role of favorable substances. Before discussing characteristics of fungitoxicants that favor penetration of fungal membranes, a description of these membranes is in order.

*Nature of Membranes.* Cytoplasmic membranes, as first described by Davson and Danielli (1943), are considered to consist of three layers: an outer and inner layer of hydrophilic materials of which protein composes the major part and a middle layer of lipids. Actually, the lipid layer is considered to consist of two layers of fatty materials lying with their polar groups oriented toward the protein layers and with their nonpolar portions intermingling with each other. ROBERTSON (1959) has proposed that phospholipids comprise the polar portions of the bimolecular leaflet in most organisms and that these are held together by VAN DER WAALS forces among the hydrocarbon chains. However, this unit membrane concept, common to most organisms, has been challenged by KORN (1966) on the basis of little experimental support for specific orientation of phospholipids or for the polar bond between lipid and protein. In any case, cell membranes appear to consist of a sandwich of proteins with a spread of a bimolecular lipoid dressing.

The fatty acid tails are largely of C 16 to C 18 fatty acid. A portion of the fatty tails is unsaturated, which, according to QUASTEL (1967), affords stability to the bimolecular sheath. Saturated fatty acids would afford porosity in the lipoid layer for the passage of water and other small hydrophilic molecules. The ratio of 1:2 of saturated to unsaturated fatty acids appears optimum for stability and porosity of the lipoid layer of membranes (QUASTEL, 1967).

The proteinaceous layers of the membrane serve many purposes. The polymer chain of peptides and the cross links of the side chains enable the protein to form a layer of structural strength. The membrane proteins can serve as enzymes to catalyze metabolic functions, of which the conversion of certain metabolites to suitable forms for passage through the membrane may be important.

*Binding to Membrane.* In studies on uptake, fungous cells are bathed in aqueous solutions of toxicant. Since the cell wall of conidia is considered a loosely knit structure, the cytoplasmic membrane would come in contact with the ambient solution. Solutes may travel quickly along the interphase between ambient solvent and membrane by surface phenomena. The outer surface of protein has numerous types of binding sites on which the toxicants can be attached. The nature of the binding sites would vary with the side chain of the amino acid residue of the protein. Bonding can proceed through electronic and polar bonds of acidic, hydroxyl and amino groups. The toxicants can bind to alkyl groups of protein through the exclusion of water molecules or hydrophobically. Recent evidence has shown that this hydrophobic bonding is one mode of attachment of small molecules to proteins (ANDERSON and REYNOLDS, 1965). The coating of the cytoplasmic membrane with fungicide may explain the rapid rate of uptake.

The surface phenomena and binding of fungicide may not be restricted to the membrane at the cell surface. Current concepts of cell morphology as reviewed by MILLER (1962) suggest that the cytoplasmic membrane is larger than that of the cell surface. Apparently, the membrane contains small penetrating invaginations or pinocytes. In addition, deep folds of membrane are believed to penetrate into the protoplast. Conceivably, the portion of the fungicide bound to these extensions of the cytoplasmic membrane may constitute a large part of the uptake.

*Passage Through Membrane.* According to MILLER (1962), compounds pass through membranes by one of five processes: mass movement through pores in the membrane, diffusion through nonselective membranes, diffusion through selectively permeable membranes, transport across membranes at the expense of energy, and engulfment through pinocytosis. Of these, diffusion through selectively permeable membranes and active transport mechanisms are most peculiar to fungicides.

Ordinarily, charged substances do not penetrate membranes of fungi in the dissociated form but permeate in the undissociated form. BARRON *et al.* (1950) found that organic acids in the undissociated form are fermented by yeast cells but their anions are not utilized. ROTHSTEIN and HAYES (1956) found no exchange of $Mn^{+2}$ and $Ca^{+2}$ ions in solution with those of the cell as measured with radioisotopes. Undoubtedly, ionic substances are converted to nonionic materials to penetrate the cytoplasmic membrane. Metal chelates that are toxic in the partially chelated form are considered to permeate fungous cells as full chelates, which are not ionized (BLOCK, 1955; ZENTMYER, RICH and HORSFALL, 1960). Free fatty acids may serve as carriers of cations across membranes, also (QUASTEL, 1967).

HANSCH and FUJITA (1964) suggested that the migration of toxicant from the ambient fluid to the site of action within the cell consists of a series of random steps of partitioning between aqueous and lipoid regions, the rate of which would depend upon the chemical and membranes. COLLANDER (1954) with cells of the alga *Nitella muscronata* found that the rate of movement of organic compounds through cellular membranes is approximately proportional to the logarithms of their oil:water partition coefficients.

Table 1. *Distribution of toxicant in fungous spores (from data of* OWENS *and* MILLER, *1957)*

| Toxicant | % distribution in *Neurospora sitophila* and *Aspergillus niger* | | | | | | | |
|---|---|---|---|---|---|---|---|---|
| | Walls | | Mitochondria | | Microsomes | | Supernatant | |
| | N.s. | A.n. | N.s. | A.n. | N.s. | A.n. | N.s. | A.n. |
| Ce | 4 | 48 | 10 | 42 | 45 | 7 | 41 | 3 |
| Ag | 4 | 18 | 25 | 29 | 45 | 14 | 26 | 39 |
| Hg | 0 | 32 | 36 | 48 | 63 | 8 | 1 | 12 |
| Cd | 0 | 48 | 14 | 35 | 35 | 6 | 51 | 11 |
| Zn | 7 | — | 10 | — | 34 | — | 49 | — |
| Glyodin | 3 | 20 | 5 | 58 | 77 | 16 | 15 | 6 |
| Dichlone | 2 | 6 | 4 | 14 | 19 | 2 | 75 | 78 |

RICH and HORSFALL (1952) showed that fungitoxicity of alkyl-nitropyrazoles to conidia of *Stemphylium sarcinaeforme* was related to the oil:water partitioning coefficients of the compounds. WELLMAN and MCCALLAN (1946) found that fungitoxicity of 2-alkylimidazolines increased, reached a maximum and decreased with increase in the number of carbons in the 2-alkyl group. Since the length of the alkyl chain is a function of oil:water partitioning, maximum toxicity occurs with the homologue possessing the specific hydrophobic property. RICH and HORSFALL considered that the specific oil:water partitioning requirement of the compound had to do with its passage through the cytoplasmic membrane.

The optimum partition coefficient for fungitoxicity suggests that hydrophobic barriers lie between the fungicide and its site of action. The lipoid layer of the cytoplasmic membrane may be the principal barrier. Other barriers are membranes and lipoprotein of mitochondria, ribosomes, nucleus and endoplasmic membranes. To pass through the membrane, the fungicide must partition from the aqueous phase on the outside of the protoplast into the lipoid phase of the membrane and then, in turn, partition into the aqueous phase within the protoplast. With other things equal, the first partitioning increases with increases in the oil:water partition coefficient of the toxicant, but a concomitant decrease occurs in the second partitioning. Thus, the maximum rate of penetration of the membrane would occur at an optimum value of oil:water partitioning. This hydrophobic property undoubtedly serves as a basis of selective permeability of membranes.

### Intracellular Migration of Toxicant to Sites of Action

Earlier concepts of cell morphology suggested that the interior of a protoplast consisted of an aqueous solution which bathes the various organelles, globules, and vacuoles. In line with this concept, toxicants were thought to have unlimited access more or less throughout the cellular fluid once they have penetrated the plasma membrane. As was discussed earlier, this concept of cell morphology is no longer tenable. In addition to forming pinocytes, the cytoplasmic membrane is believed to fold deeply into the cytoplasm with fluctuation in the foldings (MILLER, 1962). The cytoplasm is believed to have a definite endoplasmic reticulum continuous from surface membrane to nucleus. Many areas within the protoplast are not subjected to a permeation barrier and compounds have access to these "free spaces".

Once within the cellular fluid of the protoplast, the toxicant may be carried by protoplasmic streaming. HILL (1965) considers protoplasmic streaming the primary method of transport of substances from one part of a hypha to another. Undoubtedly, the toxicant is transported within the protoplast of conidia by the same mechanism. In addition to traveling alone, the toxicant may be carried within mitochondria and ribosomes that move as free entities within the protoplast.

The data of OWENS and MILLER (1957) illustrated in Table 1 show indeed that fungitoxic compounds migrate to intracellular areas and are found in substantial amounts in mitochondria, microsomes, and cellular fluid.

The sites of action appear to be enzymes of the fungous cell. These systems are located in membranes, mitochondria, ribosomes, and the nucleus and, to a lesser extent, the cellular fluid.

## Conclusions

Apparently, the migration of toxicant to sites of action consists of a series of chemical and physical reactions that involve both the electronic and the hydrophobic nature of the toxicant. With electrolytes, the migration involves a series of cationic exchanges with salt linkages and chelate ligands of enzymes. The metals may penetrate lipoid membranes in the form of simple chelates. In the undissociated state, the toxicant may bind hydrophobically with lipids in the membranes as it passes through them. Those nonelectrolytic toxicants that are analogues of metabolites may enter the cell by means of uptake processes for the metabolites. Lipophilic toxicants appear to enter and migrate through the protoplast by a series of partitionings between aqueous and lipoid regions as they move from one lipophilic area to the next.

Chapter 5

# Sites of Action

## Introduction

A fungicide inhibits a fungus by attacking one or more sites of the cell. The inhibition can be measured through its effects on growth, sporulation, infection or decay. Often it is difficult to discern the sites primarily responsible for toxicity.

Fungous cells contain a cell wall and a protoplast. Surrounding the protoplast is a cytoplasmic membrane and within the protoplast are numerous membranes connected to the exterior. Other entities contained within the protoplast are the mitochondria, ribosomes, and nucleus. Each entity carries on functions vital to the welfare of the fungous cell. When one entity is injured beyond repair, the cell as a whole is affected.

Present knowledge of the effects of fungicides on fungi may form a basis for developing working concepts of mechanism of action. In this chapter, the actions of fungicides will be discussed in view of the damage the toxicant causes to components of the cell.

## Cell Wall

### Structure

Cell walls provide the framework within which a fungus takes shape. The structural materials in most fungous genera are chitin, cellulose, and, in certain yeasts, non-cellulosic glucose. Chitin is an amino polysaccharide consisting of *N*-acetylglucose amine subunits with *B*-1,4-linkage. Cellulose is a polysaccharide of glucose subunits with *B*-1,4-linkage. Both polymers form linear molecules that occur in cell walls as fibrils. Fungous cell walls are composed of two or more laminated layers (ARONSON, 1965). The outer layer may contain melanins. Glucans, mannans, and fucans are found in cell wall materials, also. Pectin and protein make up small percentages of the dry weight of fungous cell walls. The presence of proteins suggests that a few enzymic reactions may occur at the cell surface. Traces of other carbohydrates, lipids, and minerals are found in the cell wall, too.

Synthesis of new cell walls occurs at the tips of hyphae and germ tubes. Secondary reinforcement occurs further back from the tip. In the budding of yeast and possibly in the branching of filamentous fungi, cell wall synthesis occurs as the new cell expands. In the process of budding, a sulfhydryl reductase in *Candida albicans* weakens the wall by breaking some cross-linked disulfide bonds in the proteinaceous structures, allowing protrusion of protoplasm (NICKERSON and VAN RIJ, 1949).

## Toxicants

Toxicants that act on the cell wall either cause dissolution of components or inhibit their synthesis. Bleaching agents can remove the melanin from the outer layer of the cell wall. Lysozyme is used as a tool to digest the remainder of the cell wall from the protoplast. Although a naked protoplast, when protected by an isotonic solution, is capable of synthesizing new cell wall material, the protoplast may lyse and fail to survive in a less favorable environment.

When cell wall synthesis is stopped and growth of the protoplast continues, enlargement of the protoplast stretches the cell wall and causes the cell to swell and to twist. Protoplasm leaks out at the weak spots in the wall. The distortions and swellings of tips of hyphae and germ tubes may indicate the action against cell wall synthesis.

There is strong evidence that griseofulvin inhibits synthesis of cell wall material. The curling and swelling of hyphal tips by griseofulvin, as observed by BRIAN (1960), was considered a result of inhibiting synthesis of chitinous elements of hyphal walls. The action of the antibiotic is restricted to chitinous fungi. EVANS and WHITE (1967) found that griseofulvin alters the structure of microfibrils in the cell wall. The microfibrils of treated cells were shorter and thicker than those of untreated cells in photographs of preparations taken with an electron microscope. The authors concluded that griseofulvin acts where cell wall is being synthesized — on the inner phase between cell wall and plasma membrane at the tips of hyphae. The extra deposition of material in cell walls on swollen hyphae was considered a response of the fungus to maintain integrity of the enlarged malformed cell.

However, the action of griseofulvin may not be restricted to cell wall synthesis. BLANK, TAPLIN and ROTH (1960) found in the dermatophyte *Trichophyton rubum* that the antibiotic causes shrinkage of protoplast and formation of large fat granules in addition to a malformed cell wall. In preventing pathogenesis on plant tissue by fungi, griseofulvin inhibits development and growth of haustoria, specialized tips of hyphae that penetrate host cells to obtain metabolites of the host (DEKKER and TULLENERS, 1963). Undoubtedly, these actions are not the result of stopping just the synthesis of cell wall materials. Processes of growth and differentiation must be affected as well.

The antibiotics that prevent synthesis of cell wall materials in bacteria — penicillin, vancomycin, restocetin, bacitracin, novobiocin, and *D*-cycloserine — are inhibitors of protein synthesis (REYNOLDS, 1966). These antibiotics prevent the incorporation of amino acids into the mucopeptide component of the wall. New cell walls are not synthesized, thus the naked protoplast is exposed and then lyses. The action of penicillin and D-cycloserine is restricted to cell wall synthesis. But this may be only one target for the others. Vancomycin, restocetin and bacitracin act on the cell membrane too. Novobiocin affects bacteria in a variety of ways in addition to acting on the cell wall and membrane.

HORSFALL (1956) lists 41 compounds that cause abnormalities at the tip of hyphae and germ tubes. Among these are metal salts, esters and cyclic esters of carboxy, thio, and thiono acids, phenols, amino and lipophilic compounds. HORSFALL considers the effects of these compounds on the cell wall to be indirect, *e.g.*, altering formation or inhibiting activity of enzymes for synthesis of cell wall. Cycloheximide, an inhibitor of nucleic acid and protein synthesis, and 2,4-dinitrophenol, the classic uncoupler of high energy phosphates, cause thickening of the cell walls of hyphal tips (BENT and MOORE, 1966).

Since the cell wall acts as a repository for many toxicants, the concentration of toxicant is not necessarily indicative of toxic action occurring at that site. Practically all of the griseofulvin taken up by sensitive fungi is bound to the cell wall in the tips of hyphae (BENT and MOORE, 1966). A complete toxic dose of cationic fungicides is adsorbed by cell walls in dormant ascospores of *Neurospora* (LOWERY, SUSSMAN and VON BOVENTER, 1957). Conidial walls of *Alternaria tenuis* and *Penicillium* can bind large amounts of copper. The toxic action of these cationic fungicides is considered to occur at sites other than those on the cell wall.

However, toxicants bound to the cell wall may interfere with functions at the cell surface. Fungi are known to produce enzymes that digest complex food material extracellularly. These enzymes are produced on and act near the surface of the cell. Conceivably, bound toxicants can inhibit the synthesis or activity of extracellular enzymes. For example, carbohydrase and ascorbic acid oxidase are located near the spore surface and are inactivated by an acid rinse [MANDELS, 1953 (1, 2)].

## Cell Membrane

### Structure

Unlike the weblike fibril construction of cell wall, the plasma membrane is considered to be composed of packets or subunits of membrane (SALTON, 1967). Each subunit consists of a bi-leaflet lipoid layer between two layers of protein. The subunits are joined together by metal bridges and hydrophobic bonds to form a continuous membrane (RAZIN, MOROWITZ and TERRY, 1965). Membranes in cell-free preparations can be broken down to subunits by chelators, surfactants or sonic vibration. The last two agents attack the hydrophobic bonds and the chelators attack the metal bridges. When subunits, produced by treatment with surfactants, are washed free of the chemical and treated with polyvalent cations, they reaggregate and can perform some functions of the membranes (RAZIN, MOROWITZ and TERRY, 1965).

### Toxicants

*How They Affect Passage of Solutes.* Surface active fungicides do attack membranes of intact fungous cells. Dodine causes leakage of cellular constituents from yeast cells and conidia (BROWN and SISLER, 1960; SOMERS and PRING, 1966) and glyodin destroys selective permeability to allow passage of electrolytes (KOTTKE and SISLER, 1962). Many antibiotics having surfactant properties cause leakage of cellular constituents from treated cells. Presumably, the surfactants destroy selective permeability of cellular membranes, partly by attacking the hydrophobic bonds at the junctures of subunits to cause gaps in the membrane. In addition, the saturated alkyl chains of the fungicide may form pores in the lipoid layer of the membrane, penetrating between the unsaturated fatty acids and forcing them apart.

Other lipophilic toxicants have been reported to change permeability of the cell membrane. RAMSEY, SMITH and HEIBERG (1944) concluded that the bursting of conidia of *Diplodia natalensis* by vapors of diphenyl is due to membrane damage caused by the toxicant. Diphenyl, at toxic concentrations to conidia of *Fusarium solani*, prevents uptake to phosphates (GEORGOPOULOS, ZAFIRATOS and GEORGIADIS, 1967). The toxicant stimulates the uptake of potassium ions. No leakage of cellular constituents was observed by the treatment. Diphenyl has no effect on cell membranes of a strain of *F. solani* resistant to the fungicide.

The metal bridges linking subunits may pose as a target of toxicity to membranes. MILLER and MCCALLAN (1957) have reported leakage of phosphorous materials from conidia treated with silver ion. Presumably, through cation exchange, the monovalent silver ion attacks magnesium and calcium bridges of the subunits to cause gaps in the membrane. PASSOW and ROTHSTEIN (1960) have found mercuric chloride to cause leakage of potassium ion from yeast cells. The mercuric ion has a greater affinity for membranes than for matrix of organelles (MATSUI *et al.*, 1962). Apparently, mercuric ions may attack the protein layer as well as the metal bridges to create leakage in the membrane. The denaturing of protein by reaction of mercury with the functional groups is well known and the metal, because of its size, may make the metal bridges too long. Tributyl tin naphthenate inhibits phosphorous uptake by conidia of *Rhizopus nigricans* (IWAMATO and KIKUCHI, 1963). Since the organo-tin hinders the utilization of phosphate, too, its action may not necessarily be on the membrane.

There is some evidence to indicate that chelating agents can attack the metal bridges of membranes of intact fungous cells. Sodium-*N*-methyldithiocarbamate was reported by WEDDING and KENDRICK (1959) to cause leakage of cellular constituents in *Rhizoctonia solani*. They suggested that the dithiocarbamate acts by attacking thiols of the membrane. However, it is possible that the compound removed metals from the membrane. The dithiocarbamates are well-known chelators of metals. In addition, methyl isothiocyanate, a decomposition product of *N*-methyldithiocarbamate, reacts with thiols and was without effect on the membrane. Other thiol reactants such as captan and dichlone do not act on fungous membranes (KOTTKE and SISLER, 1962, and MILLER and MCCALLAN, 1957).

In bacteria, ethylene diamine tetraacetic acid (EDTA) increases permeation of benzyl penicillin, presumably by a unique chelating property (HAMILTON-MILLER, 1965). Its action is pH dependent. EDTA did not induce lysis or cause leakage of cellular constituents. Other metal reacting compounds such as amino, thiol and alkylating agents had little effect on intact cells or cell-free suspension.

Dyrene at sublethal dosages has been reported by BURCHFIELD and STORRS (1957) to cause cellular constituents to leak from fungous spores. Initial uptake of the fungicide is as rapid as that of other organic fungicides, but when the membrane is altered, the rate of uptake approaches that of diffusion. Since dyrene is a powerful alkylating agent, it may alter membrane functions by entering secondary addition reactions with functional groups on the protein layer of the membrane. The reactions may rupture the protein superstructure of the membrane.

KOTTKE and SISLER (1962) tested the functioning of membranes of intact yeast cells by ascertaining if cells treated with fungicides would ferment pyruvate ion. A functioning membrane would not allow pyruvate to permeate to the fermenting enzymes. Cells treated with glyodin, dodine and nystatin (a polyene antibiotic) fermented pyruvate, while cells treated with cycloheximide or captan did not. The latter two fungicides are not considered membrane toxins.

The antibiotics vancomycin, vestocitin and bacitracin act on plasma membranes of bacteria. Tyrocidin ruptures cell membranes and causes loss of metabolites (HOTCHKISS, 1944). Polymixin causes leakage of metabolites, too. It binds specifically to the membrane (NEWTON, 1958). Asochytin causes membranes of several fungi to leak cellular constituents (OKU and NAKANISHI, 1966). The membranes of fungi resistant to the antibiotic are not destroyed.

With fungi, the polyene antibiotics are potent toxicants of the cell membrane. The polyenes bind to the cell membranes in an irreversible fashion. When bound, they destroy the selective permeability of the membrane. Cellular constituents leak out and the organism is unable to concentrate metabolites from the ambient environment. However, no visible injury to cellular membranes of *Candida albicans* from filipin and amphotericin B was observed by GALE (1963) with use of the electron microscope.

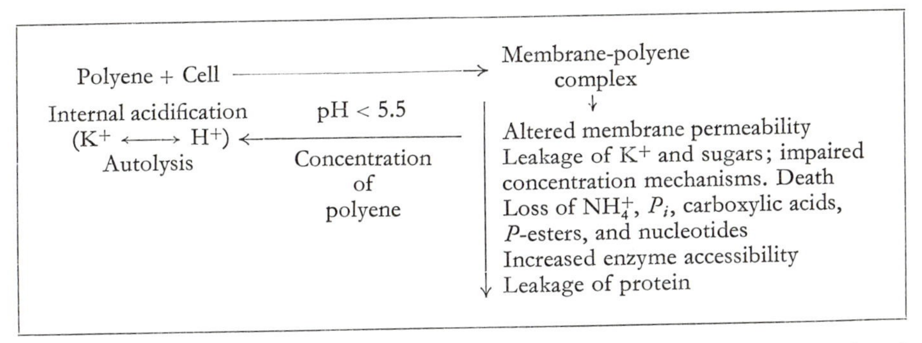


Fig. 1. Concept of polyene action. (From LAMPEN, J. O.: Interference by polyenic antifungal antibiotics (especially nystatin and filipin) with specific membrane functions. In: Biochemical Studies of Antimicrobial Drugs, pp. 111—130. Cambridge: University Press 1966)

LAMPEN's (1966) scheme for action of the polyenes is shown in Fig. 1. Swelling and lysis of fungous protoplasts is typical of polyene action.

Apparently, sterols are the materials in cell membranes that bind with the polyene antibiotics (LAMPEN, 1966). Membrane sterols are considered to be contained in the lipoid layer (VAN ZUTPHEN, VAN DENNEN and KINSKY, 1966). GOTTLIEB *et al.* (1960) found that added sterols protect fungi from action of these antibiotics and LAMPEN, ARNOW and SAFERMAN (1960) reported that sterols bind with the polyenes *in vitro*. Sensitivity of an organism to the polyene antibiotics is geared closely to the presence of sterols in its membranes. The degree of binding of nystatin is roughly proportional to the content of ergosterol in protoplasts of yeast (LAMPEN *et al.*, 1962). Bound antibiotic and sterol are released in approximately parallel proportions on disruption of membrane preparations.

The common structure of all polyenes is a macrolid with a $\beta$-hydroxylated portion and a conjugated double bond system in the lactone ring (LAMPEN, 1966). These structures give the molecule a hydrophilic portion and a rigid lipophilic portion. Other substituents to the ring such as carboxyl groups, hydrocarbon side chains, amino sugars, and epoxides are peculiar to individual antibiotics, but are not essential to the basic polyene action.

The polyenes are divided into two groups, the smaller tetraenes (having four conjugated bonds) and the larger heptaenes (having seven conjugated bonds) (LAMPEN, 1966). The smaller polyenes — pimaricin, filipin, rimocidin, estrucomycin, and fungichromin — are more toxic to fungous membranes than the larger polyenes — nystatin, candidin, candicidin, *N*-acetylcandidin, *N*-acetylcandicidin, amphotericin A and B and fungimycin. The action of the toxicants is tied more closely to the number of carbon atoms than to the number of conjugated double bonds. Although the tetraenes are better lysing agents, the heptaenes are stronger inhibitors of growth.

*How They Affect Passage of Water*. Fungous cells enlarge as the expanding protoplast presses against the flaccid cell wall. The major factor of protoplasmic expansion is the uptake of water. Mercuric salts may cause a deleterious exhaustion of water because the protoplast shrinks in the presence of these salts (HASHIOKA and TAKAMURA, 1956). Many fungicides kill by coagulating the protoplast. Apparently, the heavy metals precipitate protein.

HORSFALL (1956) proposed that fungicides can induce irreversible leakage of water from cells with subsequent death from desiccation. He suggested that water leakage could be tested by observing shrinkage of treated cells in air. He indicated that fatty acids may be toxic by this mechanism. ZARACOVITIS (1964) found that fatty acids shriveled conidia of powdery mildew pathogens with maximum shriveling from a 12-carbon fatty acid.

With conidiophores of *Alternaria solani* — a leaf spotting pathogen of tomato — maximum shriveling occurred with an 11-carbon fatty acid [HORSFALL and LUKENS, 1965 (2)]. Dodine and the 13-carbon chain analogue of glyodin also caused water loss in conidiophores of *A. solani*. Apparently, the 11-13 carbon alkyl chain is essential for the water loss action predicted previously by HORSFALL (1956).

Decanoic acid and other alkyl-chained compounds apparently trigger a runaway uptake of water by the protoplast [HORSFALL and LUKENS, 1965 (2), 1966 (1)]. Shortly after conidiophores of *Alternaria solani* are treated with these compounds, the protoplast takes up water at such a rapid pace that the hyphal tip is ruptured and cellular liquid is expelled from the tip. Following the initial rupture, liquid bleeds from several cells near the tip of the conidiophore for about 30 min until the protoplast succumbs. When this occurs, the affinity for water is past. The protoplast shrinks, the cells collapse, and the conidiophore shrivels. The treated conidiophores bleed excessively in the presence of abundant moisture. These surface-acting compounds may increase water uptake by hydrating lipophilic materials in the protoplast to cause an increase in mass.

Nonlethal collapse from dessiccation and high osmotic environment differs from the above types of cell collapse. Although naked protoplasts may die from these treatments, intact mycelium can continue to grow in a collapsed condition. Conidia can germinate while dry and produce collapsed germ tubes. Conidiophores and conidia of *A. solani* and *Helminthosporium vagans* grown by the filter paper method (LUKENS, 1960) remain viable after being forced to collapse and regain turgor several times by alternately drying and wetting the paper.

# Mitochondria

## Structure

Mitochondrial structures are the large cytoplasmic organelles which harbor the energy-generating enzymes of the cell. The structure of mitochondria has been reviewed by LINDENMEYER (1965). These organelles are delineated by a multilayered membrane 180 Å units thick. The inner layer of the membrane has infoldings that compartmentalize the interior. These internal membrane elements or "cristae" have a core of protein with two coverings of lipids. The mitochondria continually change shape and location within the cytoplasm as they carry out their enzymic functions.

## Toxicants

Little is known about the effects of fungicides on the structure of mitochondria. RICHMOND *et al.* (1967) found little alteration of mitochondria in dormant conidia of *Neurospora crassa* treated with a toxic dose of captan. However, mitochondria in treated germinating conidia failed to stain with mercury orange. Also, they were less distinguishable than unstained mitochondria of the control cells. Benzalkonium chloride causes loss of identity of mitochondria and other cytoplasmic materials in treated cells of *Candida albicans* (GALE, 1963).

As indicated in Chapter 4, mitochondria of intact cells bind fungicides. The specific binding power of mitochondria depends upon the fungous species and the fungicide. OWENS and MILLER (1957) found that the mitochondrial fraction of disintegrated conidia of *Neurospora sitophila* and *Aspergillus niger* have a high affinity for heavy metal toxicants. The cationic fungicide glyodin was peculiar. The mitochondrial fraction of *A. niger* had a high affinity for glyodin while that of *N. sitophila* had a low affinity for the fungicide. The mitochondria of both fungi had low affinity for dichlone, a chlorinated quinone. Mitochondria from conidia of *N. crassa*, *Alternaria tenius* and *Penicillium italicum* have a high affinity for cupric ion, but the binding capacity of the mitochondria varies independently with that of other cellular components between the three fungous species [SOMERS, 1963 (1)].

Most carbohydrates enter fungous cells as simple sugars. These sugars are split to 3-carbon fragments during glycolysis and then decarboxylated to 2-carbon fragments prior to entering the tricarboxylic acid cycle (TCA). In TCA, the 2-carbon fragment is oxidized to carbon dioxide and water. In the process, liberated protons are carried by flavin ($FMNH_2$ and $FADH_2$) and phosphonucleotides (NADH and NADPH), and electrons are carried by the cytochrome system. The energy is liberated stepwise and is stored in the proton acceptors and to a greater degree in high energy compounds through oxidative phosphorylation. The precise reactions involved in these processes are amply described elsewhere (COCHRANE, 1958; LINDENMAYER, 1965).

The 2-carbon fragments are also building blocks for synthesis of cellular constituents. Cofactors for numerous enzymic reactions involved in synthesis are $FMNH_2$, $FADH_2$, NADH, and NADPH. The high energy compounds, ATP and NADPH, provide the energy for synthesis and for maintaining the machinery of energy production.

Respiratory enzymes responsible for the various processes are located in various areas of the fungous cell. Enzymes that break up small molecular sugars to hexoses may be found in or on the cellular membrane. Glycolytic enzymes, especially those of fermentation, are thought to occur in cytoplasmic fluid because their activities are found in the supernatant of disintegrated cells. Enzymes of TCA, the electron transport system, and oxidative phosphorylation occur as functional units in the mitochondria. Since the activities of all of these systems are interrelated, the actions of fungicides on the energy-producing systems are discussed together under the mitochondria heading.

*How They Affect Respiration.* Since uptake of oxygen and evolution of carbon dioxide are tied directly to the energy producing systems, attacks by fungicides on any of the enzymes involved can affect respiration of fungous cells. McCallan, Miller and Weed (1954) examined the effects of 17 fungicides on respiration of conidia of five fungi. Certain of these — salts of silver, mercury, and copper, malachite green, captan, and glyodin — were strong inhibitors of respiration. Salts of cadmium, zinc, and cerium, sodium arsenate, nabam, and cycloheximide had no effect. Chloranil and dichlone, ferbam, lime sulfur and phenol were strong inhibitors of respiration of certain fungi, but had little or no effect on respiration of others.

Apparently, dodine is more toxic to anaerobic respiration than aerobic (Brown and Sisler, 1960). Both sodium methyl dithiocarbamate and methyl isothiocyanate inhibit the respiration of glucose by *Rhizoctonia solani* but they do so in different ways (Wedding and Kendrick, 1959). Domsch (1964) found that methyl arsine sulfate, mercuric chloride and captan inhibit respiration of microbes in soil. Just how these toxicants act against the energy producing systems is not too clear.

*How They Affect Glycolysis.* When anaerobic respiration is more sensitive than aerobic, the glycolytic enzymes would be the logical targets of attack. Dodine may attack the glycolytic system beyond the hexose-1,6-diphosphate step because extracts of treated yeast cells failed to utilize fructose-1,6-diphosphate. Though hexokinase of yeast is sensitive to captan (Dugger, Humphreys and Calhoun, 1959), triosephosphate dehydrogenase is considered by Montie and Sisler (1962) to be the key site for inhibition of fermentation by captan. The latter enzyme is by far the more sensitive to the fungicide *in vitro*. Captan inhibits yeast aldolase, but to a lesser degree than the dehydrogenase. Aldolase from *Sclerotinia laxa* was unaffected by captan but it was very sensitive to copper sulfate (Byrde, Martin and Nicholas, 1956).

Alkylation reactions by fungicides can stop glycolysis. In animal tissue, Holzer (1964) reported that alkylating toxicants interfere with NAD pyrophosphorylase. Interference may be directed against either enzymic activity or synthesis. Since the production of NADPH occurs in fungi, NAD phosphorylase is a logical site to explore for action of dyrene, dichlone and other alkylating fungicides.

Certain respiratory enzymes have been stimulated by toxic concentrations of fungicides. In *Sclerotinia laxa*, Byrde, Martin and Nicholas (1956) reported that activity of hexokinase is increased by copper sulfate. Lime sulfur stimulated aldolase, presumably by enriching the thiol component of the system. Activity of aldolase as well as numerous dehydrogenases depends upon active thiols in an accessory capacity.

The decarboxylation of pyruvate and the entrance of the 2-carbon remnant into the tricarboxylic acid cycle is a fungicide target cited frequently. Hochstein and Cox (1956) showed that captan interferes with cocarboxylase to inhibit the decarboxylation of pyruvate in *Fusarium roseum*. Thiamine pyrophosphate can reverse the inhibition. A build-up of pyruvate in conidia of *Neurospora sitophila* when treated with captan suggests that the metabolism of the acid in the fungus is altered by captan, too. OWENS and NOVOTNY (1959) reported, with conidia of *N. sitophila*, that captan inhibits acetate metabolism by attacking Coenzyme A. Apparently, no acetyl-CoA is formed in treated cells because the fungicide oxidizes CoA-SH to the corresponding disulfide [OWENS and BLAAK, 1960 (1)]. Dichlone causes inhibition similar to that of captan by alkylating CoA-SH.

Alkylammonium chlorides compete with $NAD^+$ for alcohol dehydrogenase in yeast (ANDERSON and REYNOLDS, 1965). Inhibition increases with increase in the number of carbons in the alkyl chain. Synthetic nicotinamides (chlorides with *N*-alkyl substitution) inhibit the enzyme competitively, too, but, as the length of the alkyl chain is increased, a mixed type of inhibition prevails. Hydrophobic bonding increases when length of the alkyl chain increases, and the attachment of inhibitor to enzyme at the $NAD^+$ site then becomes firmer and more difficult to break. With large increases in hydrophobic bonding, the inhibitors may bind to the protein at sites other than that of the $NAD^+$ attachment. This second inhibition is non-competitive.

Long chain alkyl groups are broken down in fungi through *B*-oxidation with the aid of CoA-SH. Apparently, thiol-reacting toxicants that act on CoA, such as captan and dichlone, inhibit this type of oxidation. Hence, respiration of lipid reserves is likely to be inhibited by these fungicides. Since lipids are a primary source of reserve food in spores of fungous pathogens, the prevention of fat respiration may be important in toxicity to spore germination.

*How They Affect TCA and Other Aerobic Pathways.* Among enzymes of the tricarboxylic acid cycle, fumerase from *Sclerotinia laxa* is inhibited by copper sulfate (BYRDE, MARTIN and NICHOLAS, 1956). In *Neurospora sitophila*, a buildup of the α-ketoglutarate in conidia treated with toxic concentrations of captan suggests that the keto acid dehydrogenase is inhibited (OWENS and NOVOTNY, 1959). Since thiamine pyrophosphate is a cofactor in activity of α-ketoglutarate dehydrogenase, the action of captan on this system may be the same as that against pyruvate decarboxylase.

OWENS (1960) distinguished metabolic inhibition in the tricarboxylic acid cycle of *Neurospora sitophila* by sulfur-containing fungicides. Succinate, malate, citrate and keto-acids, α-ketoglutarate and pyruvate were extracted from conidia that were exposed to toxic concentrations of fungicides. The fluctuation in concentration of organic acid with the treatment is indicative of the metabolic block imposed by the fungicides. The compounds fell into three groups depending upon the differential buildup of the organic acids with treatment. The first group, which stopped the conversion of acetate to citrate and slightly inhibited succinate oxidase, included sulfur, sodium sulfide, thiram, and ferbam. The second group — ziram, nabam, and maneb — inhibited aconitase, presumably by removing the iron cofactor. Oxine had the same effect. Methyl isothiocyanate, the sole member of the third group, had no effect on acetate-to-citrate metabolism. It did inhibit dehydrogenases of malate and succinate. The distinction between actions of isothiocyanate and nabam may explain, in part, the conclusion of WEDDING and KENDRICK (1959) that methyl isothiocyanate and sodium methyldithiocarbamate inhibit fungous respiration in different ways.

Fungi have been found to respire aerobically by several metabolic paths in addition to TCA. Notably among these are the hexose-monophosphate shunt, glucose oxidase system and glyoxylic acid shunt (COCHRANE, 1958; LINDENMAYER, 1965). In *Alternaria solani*, respiration for asexual sporulation may involve glycolic acid [HORSFALL and LUKENS, 1967 (1); LUKENS and HORSFALL, 1968].

Sporulation in *A. solani* is divided into two processes — the formation of conidiophores and the birth of a conidium on the terminal cell of the conidiophore. The second process is initiated by a change in mode of cell division from the hyphal type to budding at the tip of the conidiophore (LUKENS and HORSFALL, 1969). Glycolic acid appears to play a role in the budding process. The budding process requires a flavin-dependent metabolism [LUKENS, 1963 (1)]. Glycolic acid oxidase from leaves of tobacco is a flavoprotein (ZELITCH, 1967).

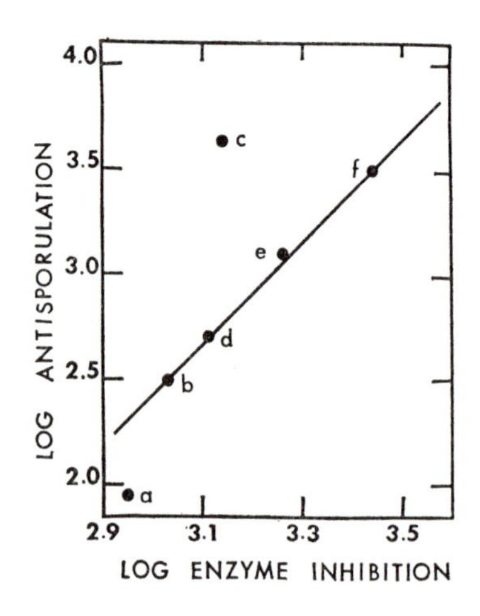


Fig. 2. Relationship of antisporulation to inhibition of glycolate oxidase from *Alternaria solani* by phenoxyacetic acids. Antisporulation is expressed as the logarithms of the reciprocal of ED-50, and enzymatic inhibition is expressed as the logarithm of the reciprocal of ED-50. The letters designate the phenoxyacetic acids; *a* 4-nitro-; *b* 2,4-dichloro-; *c* 2,4,6-trichloro-; *d* 2,3,5,6-tetrachloro-; *e* 2-chloro-4-cyclohexyl-; and *f* pentachlorophenoxyacetic acid. [From R. J. LUKENS and J. G. HORSFALL: Glycolate Oxidase, a target for antisporulants. Phytopathology **58**, 1671—1676 (1968)]

Glycolic acid oxidase can serve as a target for selective inhibition by fungicides. Phenoxyacetic acids and $\alpha$-hydroxytrichloromethanes, compounds that the fungus mistakes for glycolic acid, selectively inhibit the budding process without affecting growth of the conidiophore and inhibit the oxidation of glycolic acid to glyoxylic acid by extracts of the fungus. Glycolic acid partially reverses the antisporulant action of these toxicants (LUKENS and HORSFALL, 1968). The relationship between antisporulation and enzymic inhibition by a series of phenoxyacetic acids is illustrated in Fig. 2.

*How They Affect Terminal Oxidation.* The flow of electrons from mitochondrial NADH to the cytochrome system, a dominant pathway of respiration in certain fungi, may be a suitable target for fungicidal action. Inhibition of NADH oxidase by Dexon is concomitant with inhibition of growth of *Pythium ultimum* (TOLMSOFF, 1962).

Apparently, the action of Dexon is limited to oxidative action on NADH, because oxidative phosphorylation is not affected even at a 10-fold increase in concentration of the fungicide.

Quinones appear as likely candidates for interference with NADH oxidase. They are capable of robbing protons from the active hydrogen donor — NADH. Consequently, the quinones would be doubly toxic: (1) interfere with the flow of electrons from the NAD-cytochrome system and (2) uncouple phosphorylation when reduced to phenols as a product of the first action. The quinones can be reduced to phenols by cellular thiols. COMMAGER and JUDIS (1965) found toxicity and stimulation of respiration of glucose and succinic acid by substituted quinones concomitant in *Escherichia coli*.

Phenols are well-known uncouplers of oxidation from phosphorylation. The uncoupling is frequently accompanied by an increase in oxygen uptake by the organism. WEINBACH and GARBUS (1964) found with a series of phenols that the toxicants are bound to the protein moiety of mitochondria. They suggested that, on binding phenols, enzymes of oxidative phosphorylation assume configurational changes whereby coupling activity is lost. Proteins may require specific sites for binding phenols. From a number of purified proteins, WEINBACH and GARBUS (1964) reported that only serum albumin, myosin, cytochrome c and mitochondrial protein interact with pentachlorophenol.

In the polyphenol oxidase system of fungi, phenols are oxidized to quinones and these in turn are polymerized to melanins. RICH and HORSFALL [1954 (1)] found a good relationship between fungitoxicity to *Stemphylium sarcinaeforme* and prevention of pigmentation in fungous extracts by various quinones and phenols. Polyphenol oxidases are copper-dependent enzymes. OWENS (1954) reported that fungitoxic chelates, dithiocarbamates and their corresponding mono- and disulfides inhibit activity of polyphenol oxidase from the common mushroom *Agaricus campestris*. Presumably, the chelates form stable complexes with the enzyme through copper linkages.

## Ribosomes

### Structure

Ribosomes are small organelles that are free in the cytoplasm (ZALOKAR, 1965). They can be detected with the electron microscope but, because of their small size, little is known about their structure. They are rich in ribonuclear protein, from which their name is derived. Ribosomes are a major component in the microsomal fraction on differential centrifugation of disintegrated conidia. OWENS and MILLER (1957) and SOMERS [1963 (1)] have shown that, in addition to mitochondria, the microsomal fraction concentrates a major portion of cationic fungicides taken up by intact conidia. Apparently, the ribosomes bind cycloheximide and this binding of the antibiotic may be related to fungitoxic action (SIEGEL and SISLER, 1965). Ribosomes are the principal protein synthesizing sites in fungi. Amino acids are incorporated and proteins are quickly liberated.

## Toxicants

Recent studies of antibiotics, of Botran and of chloroneb suggest that the syntheses of protein and nucleic acids are important targets for fungitoxic action (SIEGEL and SISLER, 1964; SISLER and SIEGEL, 1967; WEBER and OGAWA, 1965; HOCK, 1968). Though the synthesis of these polymers is deeply buried in fungous metabolism, these targets may also be sites of lethal action of many of the chemically reactive fungicides. However, these chemical reactants probably affect other types of metabolism concurrently and cannot, therefore, be regarded exclusively as inhibitors of polymer synthesis. Synthesis of enzymes may be inhibited by folpet (GROVER and MOORE, 1962; GROVER, 1964). Glyodin may interfere with the metabolism of nucleic acids (WEST and WOLF, 1955). From *in vitro* studies, several isothiocyanates have been reported to inhibit protein synthesis (LEBLOVA-SVOBODOBA, 1965).

Proteins and nucleic acids are functional units of the cell. They catalyze all metabolic functions, control activity of the catalysts, and contain the genetic information. Both materials are polymers of extremely large molecular size. The subunits, arranged in precise sequences, consist of 22 amino acids in proteins and four nucleotides in nucleic acids. The general scheme of synthesis of proteins and nucleic acids is (1) activation of subunits with ATP with the help of kinases, (2) arrangement of the subunits in a correct order and (3) polymerization. A brief description of structure and synthesis is given for purposes of describing possible targets of fungicides. The information was drawn largely from PAUL (1964), who gives a vivid description of the processes. Because nucleic acids are not considered to be formed exclusively in the nucleus and because synthesis of nucleic acids and synthesis of protein are closely geared to each other, synthesis of nucleic acids or sites of toxic action are discussed together with protein synthesis.

*How They Affect Synthesis of Nucleic Acids.* DNA — desoxyribonucleic acid — a major constituent of chromosomes, is a polymer of nucleotides linked by phosphate esters between 3- and 5-positions of desoxyribose. A molecule of DNA consists of two strands of the polymer held together by hydrogen bonds between bases. The base pairs for hydrogen bonding are adenine, thymine and guanine, cystosine. During synthesis of DNA the two strands separate and each polymer serves as a template for arranging activated nucleotides. The nucleotides bind to the strands through their bases and are polymerized enzymically through esterification of their desoxyribose phosphate.

RNA — ribonucleic acid — differs from DNA in several ways. RNA contains ribose instead of desoxyribose and uracil instead of thymine. It is found in the cytoplasm as well as in the nucleus and its form varies with its function. The three main types of RNA are: messenger RNA (mRNA), transfer or soluble RNA (sRNA) and ribosomal RNA (rRNA).

Messenger RNA with its base ratio close to that of DNA is believed to be synthesized from DNA in a manner similar to that of DNA duplication. Apparently, synthesis occurs by pairing in the manner of DNA duplication except for uracil pairing with guanine. Messenger RNA serves as a template for lining up amino acids in protein synthesis.

The nature of synthesis of sRNA is obscure. DNA may serve as its template as well as some other RNA. Some alterations may occur in the polymer after its formation. It acts as a carrier of activated amino acids to the mRNA template. At the site for bonding of amino acids, sRNA terminates in the base sequence cytosine, cytosine, adenine. Pseudouridylic acid and 5-methyl cytosine are unusual bases found occasionally in sRNA. At times, part of the nucleic acid strand may fold back upon itself to give a double helix appearance. Below the fold are three unpaired bases which may serve as sites of attachment to mRNA. There are specific sRNA for carrying individual amino acids.

Ribosomal RNA makes up a large part of the ribosomes of the fungous cell. Ribosomes are about 40% nucleic acid and 60% protein. Synthesis, precise structure and functions of ribosomal RNA have yet to be elucidated.

The molecular structure of DNA is critical to its role in RNA and protein synthesis. Two groups of metabolic inhibitors bind to DNA, change its structure, and subsequently inhibit its function. The first group includes the mitomycin antibiotics (WARING, 1966). Mitomycin C forms cross-links between the complementary strands of DNA. When bound by cross-links, the strands fail to separate for template purposes in subsequent nucleic acid synthesis. The cross-linking of DNA has been invoked as a mechanism of toxicity to bacteria by the mitomycin antibiotics.

The second group of DNA binding agents are the planar compounds, such as acridine, which are bound by intercalation or the sandwiching between adjacent base-pairs of the double-strand helix of DNA. According to LERMAN (1963), unwinding of the helix of DNA produces spaces between stacked base-pairs. Flat polycyclic molecules of a binding agent can be inserted into the space between stacked base-pairs perpendicular to the helix axis. Since the depth of the planar molecule of binding agent is the same as that of a base-pair of nucleic acid, the base-pairs of DNA above and below the foreign molecule are separated by twice their normal distance. The misaligned bases of DNA cannot pair properly for subsequent synthesis of nucleic acids. A bending of the DNA molecule about 45° is necessary for insertion to occur. The altered DNA stabilizes itself by the electronic interactions between the base pairs and the foreign molecule. The planar binding agents can be inserted between base pairs of RNA, but to a lesser extent when is the case with DNA (WARING, 1966). Although protein and RNA synthesis are inhibited to a moderate degree, synthesis of DNA is inhibited strongly by acridine.

Analogues of amino acids (*e.g.*, *p*-fluorophenylalanine), purine bases (8-azaguanine) and antibiotics (cycloheximide, griseofulvin, tetracyclines, actinomycin, and others) are known inhibitors of synthesis of nucleic acids and proteins in fungi. In addition to cycloheximide, KERRIDGE (1958) found that glyodin inhibits synthesis of DNA and protein. The action of this imidazoline compound, an analogue of purine bases, is nullified in part by naturally occuring purines. WEST and WOLF (1955) found that fungitoxicity of glyodin could be reversed with guanine.

Apparently, chloroneb inhibits the synthesis of nucleic acids. At fungistatic concentrations to *Rhizoctonia solani*, HOCK (1968) has found that chloroneb inhibits 85 to 90% the incorporation of thymidine into DNA fractions of the mycelium. The inhibitory action must be in a late stage of synthesis because synthesis of nucleotides from thymidine was not affected by the toxicant. Chloroneb has less of an effect on RNA and protein. The incorporation of phenylalanine into protein and uridine into RNA were inhibited 35 to 40%.

*How They Affect Synthesis of Proteins.* Proteins are polymers of $\alpha$-amino acids. The amino acids are linked through the carboxyl group of one to the amino group of the next to form a peptide bond. The polymer chain consisting of R–CHC:O–NH– units has a planar portion –C:O–N– and a nonplanar methylene bridge. The chemical and stereo conformational properties of individual proteins lie in the number of peptides in the chain and the nature of the R-groups projecting from the methylene bridges. Some proteins serve as structural materials of the protoplast, such as membranes. Others serve as enzymes that catalyze all metabolic reactions and, at times, play a role in controlling activity of enzymes.

The steps in protein synthesis are illustrated in Fig. 3. It is not clear whether the amino acids pass through the system singly or as groups. The steps are numbered for quick reference.

*Activation of amino acids*

Step 1. Attachment of amino acid to activating enzyme with the aid of ATP, which is reduced to AMP.

Step 2. The attachment of s RNA to amino acid-AMP-activating enzyme complex.

Step 3. Release of AA-sRNA, activating enzyme, and AMP from the complex.

*Arrangement and polymerization of amino acid*

Step 4. Arrangement of AA-sRNA to specific sites on mRNA.

Step 5. Polymerization of amino acids while attached to sRNA and mRNA complex.

Step 6. Detachment of peptide from the sRNA.

Step 7. Detachment of sRNA from mRNA.

Fig. 3. Steps of Protein Synthesis

Cycloheximide acts predominantly against protein synthesis, and attacks the synthesis of DNA to a lesser degree. It has little effect on RNA, amino acid synthesis or respiration. The action against DNA may be an effect of action against protein synthesis because DNA synthesis is not inhibited *in vitro* (Siegel and Sisler, 1963). Protein synthesis *in vitro* is of the same order of sensitivity to cycloheximide as growth of cells of *Saccharomyces pastorianus* (Siegel and Sisler, 1964). Moreover, the ribosomes from resistant fungous species prevent inhibition of protein synthesis by the antibiotic when the supernatant enzymes from either susceptible or resistant fungi are employed. Other effects of the antibiotic in cellular metabolism are considered a reflection of disruption of protein synthesis.

The precise mechanism of inhibiting protein synthesis by cycloheximide is not clear. The breakdown of polyribosomes to monomers, which normally accompanies protein synthesis *in vitro*, is inhibited. Siegel and Sisler (1964) found that the antibiotic has no effect on activation of amino acids or on the transfer of activated amino acids to sRNA (Fig. 3, Steps 1 to 3). It prevents the incorporation of amino acids into ribosomal protein. Whether cycloheximide inhibits the joining of AA–sRNA to mRNA (Fig. 3, Step 4) or peptide formation (Step 5) is not clear. Wettstein, Noll and Penman (1964) suggested that protein synthesis is inhibited by cycloheximide in a way which slows the rate of breakdown of polyribosomal aggregates (possibly Steps 6 to 7, Fig. 3). The failure of polymers to break down to monomers during amino acid incorporation is in contrast to inhibition by puromycin, which hastens the release of monomers (SIEGEL and SISLER, 1964). COLOMBO, FELICETTI and BAGLIONI (1965) concluded that cycloheximide inhibited a step in protein synthesis which was required for the initial binding of the messenger to ribosomes in a polyribosomal structure and for the advancement of the ribosomes along the messenger. Other antibiotics that inhibit the incorporation of amino acids into ribosomal proteins are: blastocidin S, chloramphenicol, Kasugamysin, polyoxin, and tetracyclines (MISATO 1967; FRANKLIN, 1966; SIEGEL and SISLER, 1964).

Cells treated with cycloheximide stop growing because they can no longer synthesize protein. No other damage is apparent. Toxicity can be reversed by dialysing the cells. The arrested fungus dies because it has no way to remove the antibiotic. The action of other glutarimides, *e.g.* streptovitacin A and acetoxycycloheximide on biological systems is nonreversible (GROLLMAN, 1966). These compounds have an equatorial hydroxyl or acetoxy substitution at the C-4 position which presumably confers a secondary binding site to the antibiotics. Apparently, the secondary binding site prevents removal of antibiotic from cellular components but the site is nonessential for inhibition of protein synthesis (GROLLMAN, 1966).

Cell-free protein synthesizing systems utilizing ribosomes from a species of yeast resistant to cycloheximide are not affected. Apparently, resistance is based, in part, on the protection of the target site from toxicant by the ribosome and, in part, of the failure of resistant species to accumulate much antibiotic (WESCOTT and SISLER, 1964).

Susceptibility of fungi to griseofulvin appears tied, in part, to the binding of the antibiotic to RNA (BENT and MOORE, 1966). No binding of the antibiotic by RNA of resistant fungi was detectable. At fungitoxic concentrations, griseofulvin slightly inhibits the incorporation of $^{14}C$ into protein and nucleic acids. How this inhibition occurs is not clear.

Generally, the mechanisms for inhibiting protein synthesis are clearer than those for inhibiting nucleic acid synthesis. Actinomycin is the exception to the rule. It interferes with the ability of DNA to act as a primer for synthesis of RNA in the bacterium *Bacillus subtilis* (PERMOGOROV *et al.*, 1965). Apparently the antibiotic binds the two strands of DNA together, thereby preventing their separation for RNA synthesis. The DNA is changed physically, because actinomycin-treated DNA has a higher, broader melting point and is more readily denatured than untreated DNA.

Botran and certain isothiocyanates, which are not analogues of purines, pyrimidines, amino acids and antibiotics, inhibit protein synthesis. With Botran, WEBER and OGAWA (1965) give the characteristics of action which are typical of materials that selectively inhibit protein synthesis. They are: (1) reduction in protein content, (2) less incorporation of amino acids into proteins, (3) little effect on respiration of carbon compounds, and (4) readily available resistant strains. When fungous strains susceptible to the toxicant are treated with a sublethal dosage of Botran, the free amino acid pool increases, protein content decreases and nucleic acid content remains unchanged. In contrast, cells of a fungous strain resistant to the toxicant are not affected adversely when exposed to 1000 times as much fungicide as the sensitive cells.

The inhibition of protein synthesis in *R. arrhizus* by Botran is like that of chloramphenicol in *E. coli* in which amino acids are not incorporated into protein. Chloramphenicol acts by preventing growth of peptide on the ribosomes (VAZQUEZ, 1966).

There is some evidence that chloramphenicol inhibits peptide formation by competing with mRNA for binding sites of the bacterial ribosomes. Conceivably, Botran binds to fungous ribosomes to prevent peptide formation. The binding mechanism could be via a riboside complex with the aniline group of Botran in the manner in which higher plants convert the herbicide 3-amino-3,5-dichlorobenzoic acid to a glucoside.

Isothiocyanates were reported by LEBLOVA-SVOBODOVA (1965) to inhibit protein synthesis *in vitro*. The lack of inhibition by *N*-substituted amino acids, analogous to the isothiocyanates, discounts competition of the isothiocyanates for amino acids in the system. Apparently, the isothiocyanates, well-known chemical reactants, act by combining with thiol groups on enzymes of the system.

## Nucleus

The nucleus is the organelle responsible for hereditary functions and directs control of cellular functions. The nucleus consists of a membrane and a matrix. Embedded within the matrix are other entities, such as the nucleolus. MATSUI *et al.* (1962) have shown with electron microscopy that conidia of *Cochliobolus miyabeanus* concentrate mercury first on membranes of the nucleus and other organelles and then in the matrix of nucleus and cytoplasm. No disorganization of fine structure was detected. They suggested that mercury combined with thiol and amino groups of lipoproteins. RICHMOND *et al.* (1967) found that toxic dosages of captan caused the nuclear membrane of resting conidia of *Neurospora crassa* to wrinkle. In active conidia the internal fine structure was disorganized. Captan, a thiol reactant, does not alter functions of membranes (KOTTKE and SISLER, 1962, RICHMOND *et al.*, 1967). Apparently, it disorganizes the fine structure of the cell by attacking thiols of matrix protein, not lipoproteins of membranes. Reduction in electron density of nucleus and gaps in nuclear membrane were the first morphological evidence of injury to *Candida albicans* by thiobenzoate (GALE and MCLAIN, 1964).

## Miscellaneous Components

### Free Pools

Pools of reserve thiols and amino acids within the protoplast and substrates outside of the fungous cell are sites for indirect fungitoxic action. The soluble thiols of the cell constitute a non-specific site of action as well. Over 90% of the toxic dose of captan is consumed in reactions with thiols (RICHMOND and SOMERS, 1962). The stress placed on the fungus by the toxicant to maintain a thiol reserve for proper metabolism has been suggested as a mechanism of toxic action of captan [RICHMOND and SOMERS, 1962; OWENS and BLAAK, 1960 (1)]. Dyrene, which alkylates amino groups as well as thiols, was suggested by BURCHFIELD and STORRS (1957) to act, in part, by exhausting the amino acid pool in the fungous cell.

### Extracellular Sites

Certain fungicides may inhibit the extracellular enzymes that degrade elaborated carbon-materials. GROVER (1964) found that inhibition of pectolytic enzymes of *Sclerotium sclerotiorum* and *Botrytis allii* by the captan-related fungicides folpet and difolatan was concomitant with inhibition of growth. However, activity of *Monilinia fructicola* and *Sclerotinia laxa* was inhibited by cycloheximide and dodine, both of which had no effect on that of *S. sclerotiorum* and *B. allii* (GROVER and MOORE, 1962). Although methylolmelamine, the wrinkle-proofing material for cotton fabrics, is nontoxic to fungi in the general sense, it does prevent microbial decay of the cotton fibrils (BERARD, LEONARD and REEVES, 1961). Coating of the materials prevents access of enzymes to the fibrils. Presumably, preservatives added to foodstuffs, fabrics, and wood prevent spoilage and decay by extracellular enzymes, which are secreted by the degrading organism.

## Conclusions

Fungicides attack one or more sites to cause disorder and death of the organism. They may act directly on wall, membrane, mitochondria, microsomes, or nucleus to destroy existing structures or prevent synthesis of the same. The latter action may be through nullifying functional groups of the cell, interfering with enzymic action, or interfering with the synthesis of enzymes. Fungicides may also attack the energy producing systems of the cell or cause uncontrollable release of energy that is stored. The majority of fungicides react indiscriminately with components of the fungous cell. Their selective action toward fungi is based upon their rapid uptake and accumulation by the cell. The antibiotics and other specific toxicants inhibit specific systems and selectivity is based in part on the affinity of the toxicant for particular systems.

Chapter 6

# Reactions of Fungicides with Cellular Constituents

## Introduction

Most agricultural fungicides react with functional groups of the cell. The functional group most frequently attacked is thiol. Glutathione, a free thiol in the cellular fluid, bathes the organelles of the cell. Its presence protects other thiols. Cysteyl branches on proteins serve as functional groups for enzymic activity and determine, in part, the structural form and characteristics of the proteins. Coenzyme A, lipoic acid and possibly thiamine act through functional thiols.

The second most frequently attacked is the amino group. Asparagine, glutamine, and several amino acids comprise the amino pool in the cellular fluid. Amino groups of protein, which determine, in part, structure and enzymic activity of proteins, are targets for fungicides. Histidine, hystidyl protein, and bases of nucleic acids contain the imino group, another nitrogenous functional group that is attacked by fungicides.

Other functional groups of the cell that serve as possible sites of attack by fungicides are carboxyl, hydroxyl, and aldehydes. These groups are found also in solutes of cellular fluid and on prosthetic compounds and protein.

Protein, carbohydrate polymers, and lipids may bind fungicides in transit and, in part, at sites of action through VAN DER WAALS forces and by hydrogen bonding.

The types of reaction that fungicides enter with cellular constituents will be discussed in this chapter. Emphasis is placed on the importance of these reactions in fungitoxic mechanisms.

In chemistry, there are certain models for categorizing reactions. However, most chemical reactions differ from the model because of imperfections that compete with, interfere with, or distort the model. Hence, in biochemical studies where interference and competition are paramount, one can merely allude to the model systems for the reactions of fungicides with cellular components. Model systems identified with or proposed for reactions of fungicides fall into one or more of the following categories: nucleophilic displacement, oxidation-reduction, chelation, and complementary or accessory reactions.

## Nucleophilic Displacement Reactions

A fungicide binds covalently with a thiol group by displacing the hydrogen atom of the thiol. One nucleophil (fungicide) displaces another (hydrogen or protron).

Reaction 1. $$F + RSM \longrightarrow RSF + H$$

The reaction is known as a nucleophilic substitution (Sn). The mechanism of the displacement can be either monomolecular (Sn1) or bimolecular (Sn2). The mode of reaction depends upon fungicide, cellular group, and solvent conditions.

In the monomolecular nucleophilic displacement reaction the fungicide is first converted to an active reactant which then combines with the cellular group. The driving force or rate-determining step is the conversion of fungicide to its active form. Sn1 type reactions usually obey first order kinetics. The two-step process is:

Reaction 2. a) F ⟶ (F)
b) (F) + RSM ⟶ F—SR + M

The active form of the fungicide (F) may be either an ion or a free radical. The hydrolysis of chlorinated hydrocarbons, involving the carbonium ion (HINE, 1962), is a typical example of reaction 2, Step A. Fungicides containing reactive chlorine may give rise to carbonium ions under particular circumstances (MOJE, 1960). Elemental sulfur and thiram have been proposed to react in the form of free radicals with cellular constituents (OWENS, 1960). Thiram gives rise to free radicals on heating (TAKEBAYASHI, SHINGAKI and MIHARA, 1966). However, the presence of free radicals in reactions of the sulfur-containing compounds under physiological conditions in the aqueous environment of the cell has yet to be reported. The possibility of participation of a free radical in toxic reactions of ethylene oxide, the dry sterilant, is suggested by TAKEBAYASHI *et al.*, 1966).

In the bimolecular nucleophilic displacement reaction, a transition bimolecular intermediate of fungicide, thiol, and leaving groups is formed.

Reaction 3. F + RSM ⟶ (F—RS—M) ⟶ RSF + M.

Since the rate of reaction depends upon activity of both fungicide and thiol, the reaction is not likely to obey first order kinetics. Alkylating agents and many other thiol-reacting fungicides enter Sn2 type reactions with functional groups of the cell. Since fungicides represent diverse types of chemicals, reaction conditions of individual fungicides will be described.

## Compounds Reacting with Cellular Thiol Groups

Quinones and other $\alpha,\beta$-unsaturated ketone fungitoxicants form addition products with thiols. The replacement of the reactive hydrogen atoms with chlorine atoms speeds up the Sn2 type reactions of these compounds. OWENS and BLAAK [1960 (2)] identified the products synthesized in the reaction between dichlone with several thiols. With simple thiols, such as 4-nitrothiophenol and *n*-octylmercaptan, a double substituted product was formed.

Reaction 4.

(O, O, Cl, Cl) + 2 R−SH ⟶ (O, O, S−R, S−R) + 2 HCl

However, with the more elaborate cellular thiols, glutathione and CoA, the reaction is not complete. The major end-product is the monosubstituted product and some unreactive dichlone. The disubstituted product of Reaction 4 was found only in trace amounts.

The dithiocarbamates are unique among fungicides. Unlike that of others, their fungitoxicity increases with the ability of the compound to ionize. Presumably, reaction with cellular constituents proceeds ionically. However, the free dimethyldithiocarbamate ion does not react with thiols (OWENS and RUBINSTEIN, 1964). Thiram, the disulfide of dimethyldithiocarbamate, does react with thiols and the reaction does not require oxygen. Because of the anaerobic nature of the reaction, OWENS and RUBINSTEIN (1964) suggested that the mechanism involves free radicals. Moreover, the reaction can go by way of a polar exchange mechanism similar to that described by CAMPAIGNE, TSURUGI and MEYER (1961). The final product is the disulfide of the thiol and the dithiocarbamate ion of thiram.

Reaction 5.

a) $Me_2N{-}C(=S){-}S{-}S{-}C(=S){-}NMe_2 + RS^- \longrightarrow Me_2N{-}C(=S){-}S{-}S{-}R + Me_2N{-}C(=S){-}S^-$

b) $Me_2N{-}C(=S){-}S{-}S{-}R + RS^- \longrightarrow R{-}S{-}S{-}R + Me_2N{-}C(=S){-}S^-$

When the molar ratios are equal, the reaction can be stopped at Reaction 5a with *p*-nitrothiophenol, but not with thiophenol, CoA and glutathione. Reaction of the latter two with thiol proceeds slowly and fails to reach completion.

Although the dithiocarbamate ion fails to undergo the thiol exchange reaction, ferbam reacts to a limited extent with *p*-nitrothiophenol. OWENS and RUBINSTEIN (1964) suggested that the ferric ion catalyzes formation of free radicals from the dithiocarbamate. THORN and RICHARDSON (1962) proposed that thiram is an intermediate of the decomposition of ferbam to the dithiocarbamate ion below pH 6. Because they failed to recover thiram from decomposition of ferbam, OWENS and RUBINSTEIN (1964) suggested that the ultraviolet spectrum which THORN and RICHARDSON (1962) measured for thiram was due to dimethylamine, a decomposition product of the dithiocarbamic acid. Fungitoxic responses to thiram and ferbam are similar and differ from those of other metal salts of dimethyldithiocarbamate (OWENS and HAYES, 1964).

Methyl isothiocyanate, commercially used as a soil disinfectant, reacts with thiols to form methyldithiocarbamic acid esters.

Reaction 6. $Me{-}N{=}C{=}S + RSH \longrightarrow MeNH{-}C(=S){-}S{-}R$

Methyl isothiocyanate can be formed by the reverse of Reaction 6 in the hydrolysis of methyldithiocarbamate (Vapam) and 3,5-dimethyl-1,3,5,2H-tetrahydrothiazine-2-thione (Mylone) in moist soil (HUGHES, 1960; MUNNECKE and MARTIN, 1964). Halogenated hydrocarbon soil disinfectants prevent the hydrolysis of sodium methyldithiocarbamate by esterifying the dithiocarbamates (MILLER and LUKENS, 1966). These resulting esters may not be converted to methylisothiocyanate in the soil. It has been postulated that fungitoxicity of the ethylene-bis-dithiocarbamates proceeds by way of the corresponding diisothiocyanate (VAN DER KERK, 1959).

Captan and other R–$SCCl_3$ compounds react with thiols in a unique way. The ability of these compounds to react quickly with thiols is essential to fungitoxicity (LUKENS, RICH and HORSFALL, 1965), yet their effectiveness in controlling disease is inversely linear to the rate of reaction with thiols (LUKENS and HORSFALL, 1967). The overall rate of reaction increases with increase in pH and is controlled in part by the dissociation of thiol [OWENS and BLAAK, 1960 (2)]. However, uptake of captan by fungous spores is unaffected by pH between 2.3 to 6.5 (RICHMOND and SOMERS, 1962) and toxicity to yeast cells becomes greater as the pH of the medium is decreased [LUKENS and SISLER, 1958 (1)]. The latter biological response may be due to hydrolysis of fungicide which becomes appreciable above pH 7 (DAINES *et al.*, 1957).

The reaction proceeds in 2 steps; (1) captan reacts with thiols to produce thiophosgene and then (2) thiophosgene reacts with the remaining thiols [LUKENS and SISLER, 1958 (1); LUKENS, 1959].

Reaction 7.

$$\text{(tetrahydrophthalimide)N-SCCl}_3 + 2\,\text{R-SH} \longrightarrow \text{Cl-}\overset{\overset{\text{S}}{\|}}{\text{C}}\text{-Cl} + \text{(tetrahydrophthalimide)NH} + \text{R-S-S-R} + \text{HCl}$$

Reaction 8. $$\text{Cl—}\overset{\overset{\text{S}}{\|}}{\text{C}}\text{—Cl} + 2\,\text{R—SH} \longrightarrow \text{R—S—}\overset{\overset{\text{S}}{\|}}{\text{C}}\text{—S—R} + 2\,\text{HCl}$$

Highly reactive thiols participate in Reactions 7 and 8 in the ratio of 4 moles per mole R–$SCl_3$. The rate limiting step is Reaction 7. When less reactive thiols are involved, Reaction 7 proceeds faster than Reaction 8. In this situation, thiophosgene accumulates and is lost through vaporization and hydrolysis [LUKENS, 1963 (2); SOMERS, 1967].

The fate of the trithiocarbonate in Reaction 8 varies with the thiol. Stable trithiocarbonates are formed by thiophenols [OWENS and BLAAK, 1960 (2)]. An unstable trithiocarbonate is formed by sodium dimethyldithiocarbamate. In this reaction, carbon disulfide and thiuram monosulfide are the end products (LUKENS, 1959). In the reaction of captan with glutathione, slightly more than two moles of thiol are consumed and only oxidized glutathione is detected. No trithiocarbonate is detected [OWENS and BLAAK, 1960 (2)].

Thiophosgene combines with the amino and thiol groups of cysteine to form the cyclic 2-thiazolidinethione ring instead of a trithiocarbonate [LUKENS, 1958; LUKENS and SISLER, 1958 (2)]. It reacts with amino compounds to form thiourea derivatives. The reaction of thiophosgene with thiols is a typical example of an Sn2 type reaction. The chlorine atoms are replaced by the nucleophilic group of the other reactant.

Thiophosgene hydrolyzes in water to hydrogen sulfide and carbonyl sulfide. The second compound was suggested to be carbon disulfide because the reaction products of the two vapors with diethylamine had the same ultraviolet absorption spectrum

[LUKENS and SISLER, 1958 (1)]. However, the vapor has since been identified as carbonyl sulfide from infrared gas analysis (SOMERS, RICHMOND and PICKARD, 1967).

Two interpretations have been offered for the reaction of captan with thiols (Reaction 7). Because the reaction appears like the Sn2 type, OWENS and BLAAK [1960 (2)] suggested that a thiol displaces a chlorine atom of the trichloromethyl groups. Such an intermediate would be unstable and give rise to thiophosgene and free radicals of imide and thiol as a result of pairing of one electron from N–C bond with one from C–S bond to form the thione moiety (C= S). The free radicals would attack another molecule of thiol to form tetrahydrophthalimide and disulfide.

The second and more probable interpretation involves the N–S bond in which the imide nucleophil is exchanged by the thiol [LUKENS and SISLER, 1958 (1); OWENS and BLAAK, 1960 (2)]. The resulting mixed disulfide reacts with thiol to form trichloromethylmercaptan and disulfide. The mercaptan then decomposes to thiophosgene and hydrochloric acid.

Other evidence suggests that captan reacts with thiols by means of nucleophilic displacement. On hydrolysis of captan at the N–S bond, imide and thiophosgene are formed. The rate of this hydrolysis increases with pH and occurs rapidly in strong bases (DAINES *et al.*, 1957). When captan decomposes at 200 °C, thiophosgene is formed.

The rate of the reaction between R–$SCCl_3$ and thiols depends, in part, on the R-group (LUKENS, 1966; LUKENS, RICH and HORSFALL, 1965). High rates of reaction proceed with highly electronegative R-groups that draw electrons away from the sulfur atom. Weakly electronegative R-groups cause the reaction to proceed slowly. Reaction 7 fails to occur with R-groups that block access of the sulfur atom to thiols. When the R–S site is blocked the R–$SCCl_3$ compound reacts extremely slowly with thiols by displacement of the terminal chlorine atoms.

Difolatan, like captan, reacts rapidly with thiols (LUKENS, 1962). Again, the site of attack by thiols is considered to be the N–S bond. The reaction may proceed:

Reaction 9.

$$\text{(tetrahydrophthalimido)N-S-CCl}_2\text{-CHCl}_2 + 2\,\text{R-SH} \longrightarrow \text{Cl-C(=S)-CHCl}_2 + \text{(tetrahydrophthalimide)NH} + \text{R-S-S-R} + \text{HCl}$$

The dichloroethylthioacyl chloride so produced can react with thiols through acylating and alkylating Sn2 type reactions. It is not clear if the thioacyl chloride is involved in fungitoxic reactions. However, added thiols prevent but do not reverse toxicity. Apparently, difolatan enters reactions in addition to the oxidation of thiols. Thiols react more slowly with difolatan than with captan and the fungitoxicity of difolatan is 10-fold greater than that of captan. These characteristics suggest that difolatan is not detoxified by cellular thiols to the extent that captan is in passing through the cell to its sites of action.

Cations of heavy metals form mercaptides with thiols. Fungi enriched with thiol resist metal poisoning better than fungi of low thiol content (ASHWORTH and AMIN, 1964). Resistance to metals can be induced in fungi by adding thiols to the medium

(YODER, 1951). SHAW (1954) has found a high correlation between toxic action and stability of metal sulfides.

Metals can be precipitated out of the active phase with sulfur-containing compounds. DIMOND and STODDARD (1955) showed that toxic vapors from mercurial paints in rose greenhouses can be diminished by treating the painted surfaces with lime sulfur.

Turf injury from mercurial fungicides in summer can be avoided by mixing thiram into the spray preparation (WADSWORTH, 1960). Thiram protects fungi from mercurial fungicides, also (DAINES, BRENNAN and LEONE, 1960). Apparently, thiram reduces toxicity of mercurials through reducing availability of mercury to fungi.

## Compounds Reacting with Cellular Amino Groups

Other fungicides participate in Sn2 reactions with cellular amino compounds. Dyrene is an example. Its reactive chlorine atoms are attached to a triazine ring. These react more readily with amino compounds than with thiols (BURCHFIELD and STORRS, 1956). The rate of reaction depends upon the pH of the solvent, and is maximum at the pK values for dissociation of amino and thio groups.

Reaction 10.

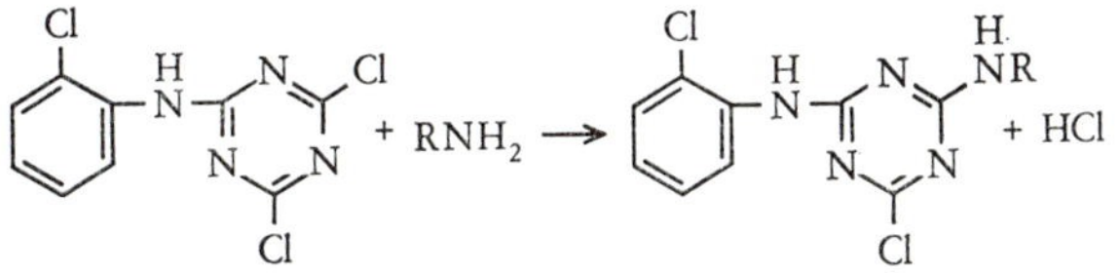

Cl, H, N, N, Cl, N, N, Cl + $RNH_2$ → Cl, H, N, N, H, NR, N, N, Cl + HCl

When the amino compound is in excess, it can react with the second chlorine atom on the S-triazine ring.

The reaction of dyrene with amino compounds is affected by the electronic effects of the amino compound on the nitrogen atom (BURCHFIELD and STORRS, 1957). Dyrene reacts slower with aspartic and glutamic acids than with their corresponding amines. However, blocking the inductive effect of the carboxyl group by insertion of methylene between the amino and carboxyl group increases activity. Hence, dyrene reacts faster with $\beta$-amino acids than with $\alpha$-amino acids. In lysine, the terminal amino group is 5 carbon atoms removed from the carboxyl group. Dyrene reacts with the terminal amino group rather than the $\alpha$-amino group. However, dyrene reacts more slowly with hydroxylysine than with lysine because of the withdrawal of electrons from the amino group by the hydroxyl group. Dyrene reacts slower with $\alpha$-amino sulfonic acids than with $\alpha$-amino carboxylic acids, presumably because of the greater withdrawal of electrons from the nitrogen atom by the sulfonic acid group. Reactions of dyrene are reduced by the steric hindrance of substituents to the amino acid compound (BURCHFIELD and STORRS, 1956). The fungicide reacts at the same rate with glycine and isoleucine, but it reacts more slowly with alanine. The methyl group of alanine blocks access of dyrene to the amino groups. A larger alkyl group, because of its electropositive nature, can overcome this steric effect, thus enhancing the reaction. The peptide bond may reduce the rate of reaction of dyrene with terminal amino groups of proteins.

Dyrene reacts with thiols at about one-fourth the rate of that with amino compounds when corrections are made for reaction constants of the thiol and for pH of the medium (BURCHFIELD and STORRS, 1958). Electronegative substituents to the thiol retard the reaction.

The position of the chlorine substitution on the aniline ring of dyrene affects the reactions of the *s*-triazine with amino and thiol compounds. According to BURCHFIELD and STORRS (1956), the order of reaction of homologues of dyrene with glycine and cysteine was *o*-chloro > *m*-chloro > *p*-chloro > no substitution. In spore germination of several fungi, the order differs slightly in that the *m*-chloro homologue is less effective than the other chloro aniline-*s*-triazine.

## Oxidation-Reduction Reactions

Strong oxidizing agents, such as hypochlorite and ozone, are well-known surface disinfectants. Also, the toxicity of the reducing agents hydrogen sulfide and sulfite is well known. Although these redox toxicants may attack and destroy functional groups of fungous cells, their action may lie in the excessive removal of electrons from or donation of electrons to the cellular system. If the excessive change in electron density cannot be equalized by the glutathione and other redox buffering systems, the cell becomes disorganized and systems responsible for growth and other metabolic functions cease to operate in an orderly fashion.

MILLER, MCCALLAN and WEED (1953) proposed that the fungitoxic action of sulfur may lie in excessive removal of protons from the protoplast. Sulfur-sensitive fungi quantitatively reduce the element to hydrogen sulfide. The source of protons for the reaction is likely to be dehydrogenase systems of respiration. On shortage of protons, these systems cannot function and energy is no longer available to the cell. In a fashion similar to the effects on dehydrogenases, sulfur may compete with oxygen for protons from terminal oxidases (HORSFALL, 1956). Interference at this point in cell respiration may stop the synthesis of high energy phosphates. The reduction of sulfur to hydrogen sulfide by fungi may involve the cytochrome system and may uncouple phosphorylation (TWEEDY, 1964).

The diversion of electrons and protons to abnormal pathways may be a toxic mechanism of dichlone in addition to that vested in its alkylating properties. Quinones may uncouple phosphorylation from respiration in fungi and they may function in this way while bound to cellular thiols (SISLER, 1963). The action is similar to that of uncoupling phosphorylation from respiration by dinitrophenol.

The toxic moiety of Karathane is probably the free phenol (KIRBY and FRICK, 1958). Esterified phenols selectively attack certain fungi. HORSFALL and LUKENS [1966 (2)] suggested that the selectivity may lie in the ability of the sensitive fungus to break the ester linkage to form the active moiety.

## Chelation

Heavy metals required in trace amounts by fungi aid enzymes in catalyzing metabolic functions. The metal may be active in this role as a chelate with the biological component.

A chelate is a cyclic structure composed of a metal atom and a ligand. At least one of the groups of the ligand shares electrons with the metal to form a coordinate bond. The donor groups are mostly nitrogen, oxygen, and sulfur atoms. Illustrated in Fig. 4 are two types of chelates of amino acids and a divalent metal ion. In one, the metal shares electrons with the nitrogen atom and binds ionically with the carboxyl oxygen, and in the other all metal bonds are coordinate.

The characteristics of the metal are altered on chelation. In the coordinate-ionic chelate, the metal is tightly held to the ligands by the multiple bonds and its positive charge is neutralized by the ionic bonds. In the chelate with coordinate bonds, the metal is firmly attached to the ligand and its positive charge is shared, in part, with the donor groups. In either case, the complex reduces the polarity of the metal ion and, in turn, increases its hydrophobic characteristics. Metal ions may pass through the lipoid layers of cellular membranes as chelates of metal and protein components of

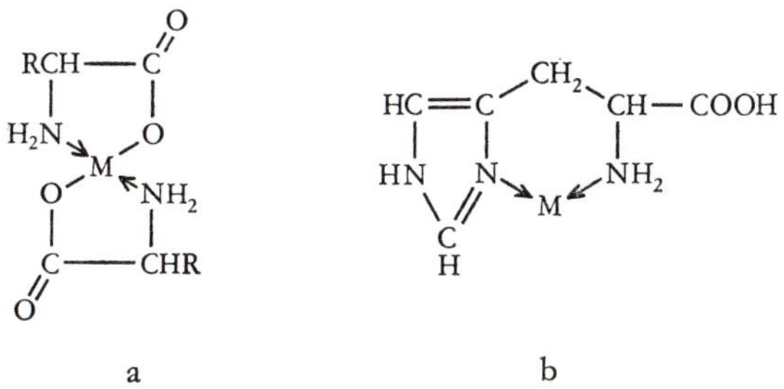

a b

Fig. 4. Chelates of amino acids and a metal: (a) a typical bidentate ligand and (b) a unique combination of metal with histidine

the membrane (Eyring, 1966). Likewise, the fungitoxic metal chelates of 8-hydroxyquinoline, the dialkyldithiocarbamates, and 2-pyridinethione-*N*-oxide are suggested to permeate fungous membranes in form of the full chelate of copper, iron or zinc (Sijpesteijn, Janssen and Dekhuijzen, 1957; Sijpesteijn and Janssen, 1958; Goksøyr, 1955; Block, 1956; Zentmyer, Rich and Horsfall, 1960; Albert, Gibson and Rubbo, 1953; and Smale, 1957).

Chelation can increase chemical reactivity of the ligand. In the metal-catalyzed hydrolysis of amino acid esters, an introduced positive charge in the vicinity of the ester increases the chance for collision of the chelate with the negatively charged hydroxyl ion (Hay, 1965). Cupric ion catalizes the hydrolysis of peptide bonds by inducing reactivity of the carbonyl moiety and by attracting hydroxyl ions to that group (Hay, 1965). The reactivity of the carbonyl moiety increases when its resonance with the amide group is quenched on coordinate bonding between the amino nitrogen and the metal ion. The subsequent withdrawal of electrons by the metal induces a positive charge in the amide, which is attracted to hydroxyl ions. Barring ionization of the proton from the amide group, hydrolysis proceeds. The catalytic activity of cupric ion is restricted to the narrow pH range between chelate formation and ionization of the peptide hydrogen.

The extent of the induced positive charge in the ligand by the metal ion depends, in part, on the metal. Both rate constants and stability constants of transition metal oxaloacetates increase in the order of Ca < Mn < Co < Zn < Ni < Cu (HAY, 1965). Apparently, metal catalysts act by forcing the ligands to their transition state for reactivity.

Metals may be toxic in the form of chelates. HORSFALL (1956) found that a parallel relationship exists between fungitoxicity and stability constants of metal complexes. The bound metal may block enzymic functions of the cell or the bound metal may catalize toxic reactions among cellular constituents. Metal toxicity brought about by blocking enzymic functions is indicated by the fungistatic nature of toxicity. Metals can be removed from fungous cells by placing them in water (BODNAR and TERENYI, 1932), dilute acid (PRÉVOST, 1807), and solutions of chelating agents (MÜLLER and BIEDERMANN, 1952).

Catalytic action requires only a small amount of the catalyst. Growth of yeast cells and growth of mycelium of *Curvularia lunata* are inhibited by doses of mercuric chloride and phenyl mercury acetate with a 3- to 10-fold increase over the threshold doses (unpublished data). The very steep slopes of the dosage response curves of the mercury salts may indeed indicate toxicity by catalytic action. Metals can catalize reactions of fungicides. Manganese causes the conversion of nabam to ethylene thiuram monosulfide, which is considered a step in the toxicity of nabam (LUDWIG, THORN and UNWIN, 1955).

Fungitoxicity by poisonous chelating agents has been proposed to proceed by one of several mechanisms. The toxicant may rob essential metals from the cell (ZENTMYER, 1943; ALBERT, 1944). With 8-hydroxyquinoline, fungitoxicity is lost when the compound is prevented from combining with metals of the cell. ZENTMYER (1944) destroyed fungitoxicity of 8-hydroxyquinoline by acidifying the medium and by precipitating the compound in the medium with zinc ion. BLOCK (1956) has since reversed the action of chelating agents by adding excess metal ions to the medium. Analogues of 8-hydroxyquinoline that do not chelate metal ions have little toxicity (ALBERT *et al.*, 1953; BLOCK, 1955).

Metal deficiency induced in fungi by toxic chelating agents supports the metal-robbing theory. In *Aspergillus niger*, dimethylglyoxine causes the fungus to grow in clumps on agar surfaces and drastically reduces its sporulation (RICH and HORSFALL, 1948). The infusion of chelating agents into humans increases the excretion of zinc in the urine (SPENCER and ROSOFF, 1966). OWENS (1960) found aconitase ($Fe^{++}$-dependent) in conidia of *Neurospora sitophila* to be inhibited by growth-inhibiting concentrations of 8-hydroxyquinoline. BLOCK (1955) reported that a fungous creosolase requiring copper is inhibited by 8-hydroxyquinoline. The work of ESPOSITO and FLETCHER (1961) with microspores of *Fusarium oxysporium* suggests that 8-hydroxyquinoline interferes in pteridine biosynthesis in the copper-mediated conversion of purine precursors to 4,5-diaminopyrimidine. The compound 4-aminofolic acid, which attacks the conversion in another way, acts in synergism with the fungitoxic chelator on the spores.

However, the metal-robbing theory may yield to one of metal poisoning for fungitoxic chelating agents. MASON (1948) found that copper 8-hydroxyquinoline is more fungitoxic than the chelating agent alone. Reducing the amount of metal chelate formed by growing fungi on metal-depleted medium or by adding nontoxic chelating agents to the medium reduces activity of the toxic chelating agent (ALBERT, GIBSON and RUBBO, 1953; SMALE, 1957; ZENTMYER and RICH, 1956).

It is commonly accepted that the half-copper chelate is the principal toxic moiety of fungitoxic chelating agents. LOWE and PHILLIPS (1962) found that copper is inserted into porphyrin rings via the half-chelate of several toxic chelates. The action of copper chelates may be the insertion of copper in natural systems in which less active metal catalysts are normally found. RICH (1960) suggested that the half-chelate of copper may be toxic by binding to enzyme surfaces to form mixed chelates with the protein. The presence of the metal chelate may tie up functional groups or may change the structural nature of the enzyme so it loses control of its catalytic activities.

Metal catalysts may act by binding cofactors to sites on the enzyme. Fungitoxic chelators may compete with the natural cofactors to bind through metal to sites on enzymes. Once bound to the enzyme, the fungitoxic chelator may not undergo change or be released from the enzyme in the usual course of enzymic activity. The inhibition can be reversed by a buildup of the competing substrate. The reversal action may be limited to the particular substrate or to compounds closely related in structure to the substrate. ESPOSITO and FLETCHER (1961) reversed the toxic action of copper-8-hydroxyquinoline in *Fusarium oxysporium* with pteridines. Only certain pyrimidines and their precursors were effective competitors of the toxicant, however.

The toxicity of chelate fungicides has been clouded by the role played by trace element nutrition in the action of the toxicants. It is compounded further by the peculiar permeation of the compounds as expressed in the unusual polymodal dosage-response curves of the chelate toxicants (DIMOND, HORSFALL, HEUBERGER and STODDARD, 1941; GOKSØYR, 1955; BLOCK, 1955; SIJPESTEIJN, JANSSEN and DEKHUIJZEN, 1957). Originally, DIMOND, HORSFALL, HEUBERGER and STODDARD (1941) suggested that the two positive slopes of the dosage-response curve represented two distinct toxicants. The negative slope of the dosage-response curve represented the change to the second toxicant at the expense of the first as the dosage of the chelator was increased. With dimethyldithiocarbamate (DDC) GOKSØYR (1955) suggested that the first toxicant was the 1:1 copper:DDC complex formed by DDC combining with copper in the medium. As the dosage of DDC was increased, the 1:2 copper: DDC complex formed at the expense of the 1:1 complex. The consumption of DDC to form the nontoxic 1:2 complex caused potency to fall, producing a negative slope. When all of the copper in the medium was consumed in the 1:2 complex, DDC by itself or in 1:1 complexes with other metals was considered the second toxicant.

In view of more recent work on 8-hydroxyquinoline, the polymodal dosage-response curve was explained in terms of the equilibria of chelator and copper complexes within the cell. The 1:1 complex formed outside of the cell permeates poorly because of its polar nature. Hence, the permeating species are the 1:2 complex and the free chelator. According to RICH (1960), the first positive slope may represent the dissociation of the 1:2 complex to the 1:1 inside of the cell. As more of the free chelator is taken up by the cell, the equilibrium shifts back to the 1:2 complex. The cut-off of the 1:1 complex intracellularly by increase of the free chelator causes toxicity to fall, hence the negative slope of the dosage-response curve. The internal dosage of the 1:1 complex is limited by availability of metal, free chelator, and chelators of the cell, while that of the free chelator depends upon the applied dosage of the compound. The 1:2 complex, having satisfied its full complement of bonding between metal ion and ligand, is nontoxic.

## Complementary Reactions

Enroute to the center of action and attachment to the site, fungicides alternately pass through a series of aqueous and lipoid stages with the aid of hydrophilic and hydrophobic bonds.

Hydrophilic bonding, like aqueous solvation, comes about as water molecules orient with the hydrogen or oxygen atoms toward the fungicides and become fixed through hydrogen bonding. In the same fashion, water molecules wet protein, other polymers, and solutes of the cell. The affinity of the fungicide and cellular components for water molecules depends, in part, on their polar properties. Hydrophilic bonding then enables the fungicide and cellular components to approach each other and enter into chemical reactions. Water is the universal solvent for most biological reactions.

In contrast to hydrophilic bonding, hydrophobic bonding takes place in the absence or exclusion of water molecules. That is, compounds or portions of compounds that assume a nonpolar nature lack the polar orientation and fail to attract water molecules. As a consequence, a water vacuum surrounds these nonpolar particles in aqueous systems, thus enabling those particles to come close together. In close association, the particles can adhere to one another through VAN DER WAALS forces.

The strength of the hydrophobic bond of a compound is determined in part by the degree of repulsion of water molecules and by the number of sites associated through VAN DER WAALS forces. Hydrophobic bonding is an associated free energy exchange phenomen of the compound and a relative measure of the property can be obtained through oil:water partitioning. FUJITA, IWASA and HANSCH (1964) have introduced the substitution constant $\pi$ as a relative measure of hydrophobic bonding (see Chapter 8). With hydrocarbon compounds the number of methylene groups in the molecule gives a relative measure of hydrophobic bonding between homologous compounds. The exchange in free energy on the binding of alkylammonium chlorides and *N*-alkylnicotinamide chlorides to yeast alcohol dehydrogenase was computed to be 0.41 and 0.37 Kcal/mol per methylene group, respectively (ANDERSON and REYNOLDS, 1965).

## Conclusions

Fungicides are two types: those that react indiscriminately with functional groups of the cell and those that enter a few reactions with cellular constituents. The majority of fungicides are of the first group. Reactions are of four general types: oxidation-reduction, secondary substitution, chelation, and complementary. Among the cellular reactants are thiol, amino, keto, acidic, chelators and metal-containing compounds.

In contrast to the general reactant fungicides, highly specific toxicants react selectively with few cellular constituents. Often, the compounds are recovered intact from treated fungous cells. The toxicants compete with substrate for sites on enzymes to inhibit enzymic activity. They may cause faulty synthesis of proteins and nucleic acids. The reactions of the specific toxicants will be discussed in Chapter 7.

Chapter 7

# Effects of Fungicides on Enzymes

## Introduction

Enzymes are often the primary sites of attack by fungicides. Information about the action of fungicides on enzymes has been limited by knowledge of fungous metabolism and techniques of enzymology. Because enzymes representative of particular functional groups showed little relationship between inhibition of a particular type of enzyme and fungitoxicity [Owens, 1953 (1)], a search for enzymic blocks in fungous metabolism that are concomitant with toxicity had to be employed. Numerous workers have shown that respiration and, in turn, activity of respiratory enzymes were inhibited by toxic concentrations of fungicides. Although respiration is necessary for growth, it is not clear whether or not these actions are causes of or are secondary to inhibition of growth. Recent attention has been focused on the inhibition of protein and nucleic acid synthesis in fungi. Increase in protein content is related directly to increases in protoplasm. The discovery of pathways of synthesis for these compounds has made these studies possible and contributed to the knowledge of fungicide effects on enzymes.

Fungicides may attack the cellular mechanisms that regulate enzymes. Enzymic synthesis and activity can be increased or decreased. The action is highly specific and usually requires a precise chemical structure of the toxicant.

Fungicides attack directly by reacting with the enzyme. Enzymic activity is mostly reduced. The action may be nonspecific when reactions of the toxicant are restricted to a particular functional group on the co- or apoenzyme. In contrast, action is highly specific when a particular structure is required of the toxicant. Thiol-reacting fungicides combine with thiols on coenzymes and apoenzymes to prevent functioning of the system [Owens, 1953 (1); Owens and Blaak, 1960 (1) and (2)]. Chelating agents may remove metallic cofactors from the system (Zentmyer, 1944). These inhibitions are of a general type in contrast to the highly specific type of *N*-alkylnicotinamide which binds to alcohol dehydrogenase, the usual site of DPNH binding (Anderson and Reynolds, 1965).

Studies of the effects of toxicants on enzymes of fungi and bacteria reveal several mechanisms of action. Although few examples of certain recently defined mechanisms can be found, the discussion of these mechanisms may suggest new avenues of search for improved fungicides.

## How Fungicides Affect Enzyme Synthesis

### Increasing Synthesis

*Inducible Enzymes.* BYRDE, MARTIN and NICHOLAS (1956) found that several enzymes of *Sclerotinia laxa* were markedly stimulated when the fungus was grown in the presence of sub-lethal dosages of captan, *o*-phenylphenol, copper sulfate, and lime-sulfur. Diphosphopyridine nucleotide oxidase was stimulated by captan and copper, hexokinase by copper, aldolase by lime-sulfur, and cytochrome c oxidase by *o*-phenylphenol. The stimulation, based upon an increase in specific activity per unit of protein, may suggest an increase in enzyme synthesis. However, the possibility that the stimulation may be directed at the activity of the enzyme is not excluded. The fungicides may stimulate enzyme activity by attacking natural inhibitors of the enzymes in the fungous cell.

The formation of penicillinase in *Subtilis* and *Staphylococcus* bacteria in the presence of high concentrations of penicillin is a clear example of enzyme induction by a toxicant (DUTHIE, 1944; and KIRBY, 1944, respectively). Synthesis of penicillinase is an aerobic process and is inhibited by respiratory inhibitors, inhibitors of protein synthesis, and toxic chelating agents (GERONIMUS and COHEN, 1957). Resistance of these bacteria to penicillin has been attributed to the production of penicillinase, which degrades the antibiotic.

*Faulty Enzymes.* By mistaken identity, fungi may incorporate a foreign amino acid, purine, or pyrimidine into proteins and nucleic acids. If the foreign unit happens to fall within the active site of the enzyme, the enzyme cannot function. Likewise, a faulty base or misalignment of bases in nucleic acid components involved in protein synthesis will alter growth. In addition, a faulty link in the protein can cause allosteric effects in the enzyme to prevent the enzyme from assuming an activated shape.

The insertion of copper by fungitoxic chelating agents into porphyrin rings where iron is normally found may be considered a mechanism of faulty synthesis. LOWE and PHILLIPS (1962) found that 8-hydroxyquinoline, sodium diethyldithiocarbamate, and 2-hydroxypyridine-*N*-oxide, all highly fungitoxic chelates, were very effective carriers of copper to mesoporphyrin dimethyl esters.

The chemotherapeutic action of ethionine against *Phytophthora cinnamoni* is antagonized when methionine is added to the system (MOJE, KENDRICK and ZENTMYER, 1963). Presumably, ethionine can inhibit growth and development of the fungus in plant tissue, in part, by faulty synthesis of coenzymes. Ethionine can form adenosyl ethionine (SHIVE and SKINNER, 1963) and, hence, block transfer of methyl groups.

### Repressing Enzyme Synthesis

The metabolism of microorganisms is under some type of coordinate control. Elaborated carbohydrates are not degraded when the organism is fed simple sugars, and compounds required for the formation of cellular components are not synthesized when the organism is fed these compounds. Enzyme repression describes a relative decrease in synthesis of a particular enzyme when the organism is exposed to a given substance (VOGEL, 1957). That is, the rate of synthesis of that enzyme is curtailed in the presence of continued protein synthesis. During normal growth, repression is a general regulatory mechanism for metabolism (MAAS and McFALL, 1964). A given repressor may control synthesis of one or more metabolically related enzymes. The repressor is considered to prevent the production of the mRNA involved in this synthesis of the particular enzyme (MAAS and McFALL, 1964).

A normal metabolite in high concentrations or an antimetabolite can act as a repressor of enzyme synthesis to inhibit cellular functions. The action of a repressor toxicant is highly specific. A single enzyme, or a group of related enzymes, is not synthesized while general protein synthesis is not immediately affected. The repressor mechanism is distinguished from that of inhibitors of protein synthesis where growth is directly affected. Enzyme repressing toxicants are one of two types: repressors of catabolic enzymes and repressors of biosynthetic enzymes (MAAS and McFALL, 1964).

*Catabolic Enzymes.* Many fungi produce little enzyme to degrade complex carbohydrates when simple carbon sources are present. In fact, there is a general sequence of utilization from simple to complex carbohydrate in mixed media, the switch to the more complex food occurring when the simpler material is exhausted (COCHRANE, 1958).

Pathogenic fungi may attack cells of higher plants by degrading pectin, cellulose, and other structural materials of the host. Certain plant pathogens attack host tissue of low sugar content (HORSFALL and DIMOND, 1957). Available host sugars may repress fungous synthesis of degradative enzymes in plants that resist those fungi. PATIL and DIMOND [1968 (1)] were able to repress synthesis of certain pectinases with glucose in the plant pathogen *Fusarium oxysporium-lycopersici*. In *Verticillium albo-atrum*, phenols repress the synthesis of the pectinases [PATIL and DIMOND, 1968 (2)]. *Helminthosporium vagans* attacks Kentucky bluegrass (*Poa pratensis*) when the grass leaves are low in reducing sugars, and the application of glucose to the turf reduces the rate of disease development for a short time thereafter (LUKENS, 1968). Prolonged treatments with sugar, however, allow the pathogen to grow on the sugar outside of the plant, finally increasing the severity of disease. Thus, a repressor for disease control must repress synthesis of degradative enzymes of the pathogen without supporting its growth. Administering substances to the host to repress synthesis of catabolic enzymes is a promising mechanism for chemotherapy.

*Biosynthetic Enzymes.* Repression of biosynthetic enzymes has been described from investigations with *Escherichia coli*. The biosynthesis of methionine, tryptophan, arginine, and pyrimidine can be repressed when the bacterium is grown on media containing these metabolites (MAAS and McFALL, 1964). The process has been called "end-product repression". As revealed with the use of certain mutants of *E. coli*, the action of the repressing agent is against the regulatory gene or mRNA and not against the synthesis of the synthetic enzyme.

Precise pathways of the syntheses of several amino acids, pyrimidines, and other cellular components have been disclosed in *Neurospora crassa* with the use of deficiency mutants. ESSER and KUENEN (1967) have listed 18 fungi, in addition to *N. crassa*, in which mutants have been found deficient in capacity to synthesize one or more biochemicals. Subsequently, with the elucidation of metabolic synthesis and development of the enzymic steps, more end-product repressing systems may be found.

The final step in the formation of proline by *N. crassa* is the reduction of pyrroline-5-carboxylate to proline. YURA and VOGEL (1959) have isolated the enzyme involved. The specific activity of pyrroline-5-carboxylate reductase varies slightly with the stage of growth and conditions of the medium. Specific activity is drastically reduced when *N. crassa* is grown on medium containing proline. In a proline-requiring mutant, specific activity of the reductase is 0.2% that of the wild-type strain grown in the absence of proline.

The mechanism of enzyme repression by an antimetabolite may proceed as follows: A structural analogue of an amino acid or another required biochemical is mistaken by the fungus for the native compound. First the synthesis of enzymes of the biosynthetic pathway is repressed. If the antimetabolite cannot be used in place of the biochemical, the fungus is adversely affected. Growth stops when the enzymes in question are diluted and when the supply of the biochemical is depleted. Thus, with enzyme repression, the specific activities of the enzymes affected are reduced and the organism fails to build up a supply of required substrate (RICHMOND, 1966). Conceivably, antimetabolites of amino acids, purines, and pyrimidines may act by end-product repression.

## How Fungicides Affect Enzyme Activity

### Increasing Activity

The activity of enzymes of intermediary metabolism is under regulatory control of the cell. Reactions of a substrate are controlled, in part, by inhibiting activity of an enzyme or enzymes at particular sites of alternative reactions. There are no reported examples of fungicides interfering with the regulatory systems to activate arrested enzymes.

Heavy metal toxicants, presumably through a role of a coenzyme, can stimulate activity of metal-dependent enzymes. OWENS [1953 (2)] found that salts of zinc, copper, and a copper, zinc, chromium complex increase activity in vitro of a polyphenol oxidase. The enzyme is copper dependent. Stimulation of metal-dependent enzymes by other metals is not universal, however, because the activity of catalase, an iron-dependent enzyme, is not affected by the metal salts.

The response of fungi to sulfur suggests that the toxicant can stimulate enzyme activity. The reduction of sulfur to hydrogen sulfide by fungi (MILLER, MCCALLAN and WEED, 1953) is a metabolic process that diverts protons from steps of normal hydrogenation reactions. The reduction is aerobic and involves a cytochrome system, possibly cytochrome b (TWEEDY, 1964). Presumably, sulfur serves as an acceptor of hydrogen atoms from dehydrogenases and, hence, activates to full capacity the enzymic steps preceding sulfur reduction.

The hydrolysis of synthetic esters by fungi is an example of stimulation of enzymic activity by providing a substrate for a non-active enzyme (BYRDE, MARTIN and NICHOLAS, 1956; WOODCOCK, 1967).

### Inhibiting Activity

A toxicant can inhibit enzymic activity by reacting with the apoenzyme, coenzyme, or substrate. As a consequence of reacting with the toxicant, the catalyst or substrate can no longer participate in the enzymic reaction. False feed-back inhibition is a special case of apoenzyme modification and will be discussed separately.

*Feed-back Inhibition.* As was discussed previously, fungi can shut down synthesis of a required product when that product is provided for the organism. A large amount of end-product may inhibit one or more enzymes on its route of synthesis. Usually, the first enzyme of the synthesis is affected. The inhibition of a biosynthetic pathway can occur by either repressing synthesis or inhibiting activity of the enzyme. Each type of inhibition operates independently, although an inhibitor may act by repression, by inhibition, or by both mechanisms. Effects of enzyme repression are expressed gradually over a period of time and effects of feed-back inhibition are expressed immediately. With feed-back inhibition, the synthesis of the enzyme is not affected, thus it is distinguished from enzyme repression.

The enzymes that affect activity of several steps in a biosynthetic pathway are affected by feed-back inhibition. The activity or inhibition of regulatory enzymes requires the binding of a substance (effector) to the apoenzyme. The end-product may be the natural effector for feed-back inhibition. Allosteric effectors are those that bind to apoenzyme outside of the reaction site to cause a steric interaction of the protein for enzymic activity (STADTMAN, 1966).

Antimetabolites of end-products may pose as artificial effectors to cause false feed-back inhibition. The compound 2-thiazole-alanine inhibits growth of *E. coli* by preventing synthesis of histidine through false feed-back inhibition (MOYED, 1961). Toxicity of 6-methyl tryptophan is related to the compound-inhibiting enzymes of tryptophan synthesis. Other toxicants may act by combining with regulatory enzymes at effector sites in a competitive, non-competitive, or mixed manner. False feed-back inhibition is a highly selective toxic mechanism of enzymic inhibition that may be exploitable to control fungous growth.

*Apoenzyme Modification.* An enzymic protein contains a special site, the reaction center, where cofactors and substrate are bound for catalyzing the reaction. In addition, functional groups of the protein outside of the reaction center may aid in binding the components to the protein or function independently in catalysis. Whether the fungicide attacks the enzyme in the reaction center or outside of that center determines the nature of enzymic inhibition.

Structural analogues compete with the substrate or coenzyme for sites in the reaction center. Enzymic activity is inhibited if the toxicant does not enter the reaction and is not removed from the enzyme. The inhibition is competitive when it decreases with increase in concentration of the metabolite. The competitive effect, as demonstrated by plots of the reciprocal of velocity of the reaction by the reciprocal of concentration of the toxicant (LINEWEAVER and BURK, 1934), gives increasing values for the slopes of the dosage-response curves with increase in concentration of the substrate. Competitive inhibition of glycolate oxidase from *Alternaria solani* by 2,4-dichlorophenoxyacetic acid is illustrated in Fig. 5 A. Apparently, the phenoxyacetic acid competes with glycolic acid for sites on the enzyme. Alkyl quaternium ammonium salts and *N*-alkylnicotinamide inhibit alcohol dehydrogenase in yeast by competing with $NAD^+$, the coenzyme of the system (ANDERSON and REYNOLDS, 1965). Heavy metal fungicides compete with metallic cofactors for apoenzymes to inhibit enzymic activity [OWENS, 1953 (1); OWENS and HAYES, 1964].

Apparently, the potency of competitive inhibitor depends, in part, upon the stability of the inhibitor-enzyme complex. The ED-50 value for enzymic inhibition is the reciprocal of the stability constant of the enzyme inhibitor complex. The complex may be stabilized further by hydrophobic bonding between enzyme and inhibitor (ANDERSON and REYNOLDS, 1965; LUKENS and HORSFALL, 1968). The relationship between enzyme inhibition and hydrophobic bonding of phenoxyacetic acids is illustrated in Fig. 6. HALL *et al.* (1960) suggested that an increase in electron attracting substituents to the benzene ring of tryptophan increases the affinity of those compounds to tryptophanase of *E. coli.* Tryptophanase can act upon tryptophan, increasing

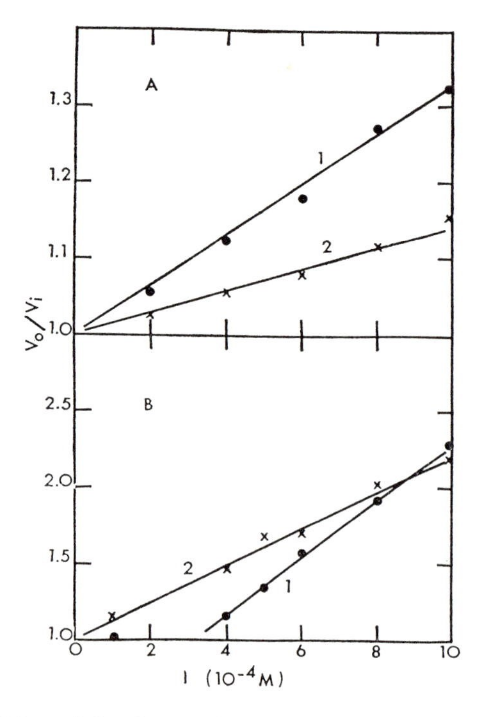


Fig. 5 A and B. Inhibition of glycolate oxidase from *Alternaria solani* by phenoxyacetic acids. A) Competitive inhibition by 2,4-dichlorophenoxyacetic acid with glycolate $10^{-3}$ M (Curve 1) and $5 \times 10^{-3}$ M (Curve 2). Mixed inhibition by pentachlorophenoxyacetic acid with glycolate, $2.5 \times 10^{-4}$ M (Curve 1) and $5 \times 10^{-4}$ M (Curve 2). $V_i$ designates the rate of glycolate oxidation 30 min after mixing enzyme, substrate and inhibitor and $V_0$ the same without inhibitor. $I$ is the concentration of the inhibitor. [From R. J. LUKENS and J. G. HORSFALL: Glycolate oxidase, a target for antisporulants. Phytopathology **58**, 1671—1676 (1968)]

activity with the addition of chloro-substituents to the benzene ring of tryptophan. An increase in electronegativity of the compound may increase the stability of the compound-substrate complex. However, an increase in length of the alkyl chain of quaternium ammonium salts increases their potency for enzymic inhibition (ANDERSON and REYNOLDS, 1965). Hence, hydrophobic bonding may play a greater role in the attachment of substrate to enzyme than does electronegativity.

Although competitive inhibition of enzymic activity can be demonstrated *in vitro*, and reversal of toxicity *in vivo*, the studies do not distinguish precisely the mechanism of inhibition (STADTMAN, 1966). When the toxicant competes with the substrate for the active site to inhibit enzymic activity, *in vivo*, the substrate may accumulate if enzymic steps involved in the synthesis of the substrate are not inhibited, also. A large quantity of substrate can force the inhibitor off the enzyme to reverse toxicity. Persistent toxicity that displays a competitive nature of enzymic inhibition may proceed by way of end-product repression or false feedback inhibition. In these mechanisms, the first enzyme of the biosynthetic pathway is inhibited and the end-product does not accumulate (RICHMOND, 1966).

Although the inhibitor mimics the substrate on binding with enzyme, it can possess other features that prevent the substrate from forcing it off the enzyme.

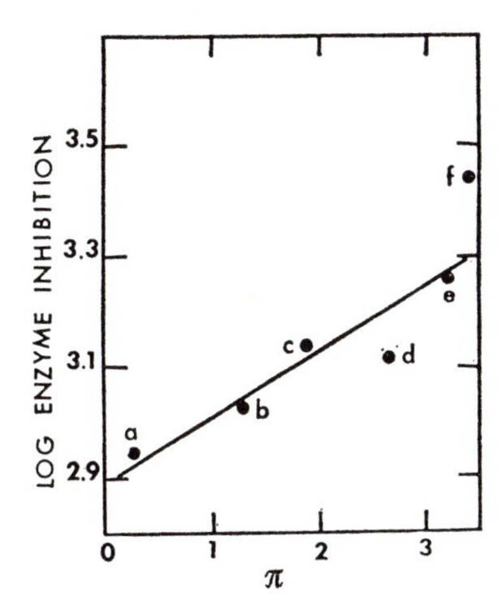


Fig. 6. Relation of inhibition of glycolate oxidase from *Alternaria solani* to hydrophobic bonding of phenoxyacetic acids. Enzymatic inhibition is expressed as the logarithm of the reciprocal of ED-50 and hydrophobic bonding $\pi$, of FUJITA *et al.* Designation of compounds is the same as that in Fig. 2. [From R. J. LUKENS and J. G. HORSFALL: Glycolate oxidase, a target for antisporulants. Phytopathology **58**, 1671—1676 (1968)]

Substituents to the inhibitor may block access of substrate to enzymic sites or may form additional bonds with the enzyme in or adjacent to the active site. Inhibition would change from competitive to mixed and to noncompetitive with increase in auxiliary bonding as an analogous series of inhibitors is ascended. An increase in chlorination of phenoxyacetic acids increases hydrophobic bonding and changes their action from competitive to mixed inhibition of glycolate oxidase (compare B with A of Fig. 5). Likewise, with yeast alcohol dehydrogenase, ANDERSON and REYNOLDS (1965) reported that inhibition by *N*-alkylnicotinamides shifts from competitive to a mixture of competitive and noncompetitive inhibition on increase in length of the alkyl substitutent.

When the fungicide binds to functional groups in or out of the action site of the enzyme, inhibition is independent of substrate concentration and is therefore non-competitive. The presence of substrate has little influence on the binding of the inhibitor. The inhibitor may block access of substrate to enzyme or prevent catalysis.

Amino-reacting fungicides inhibit amino-dependent enzymes, thiol-reacting fungicides inhibit thiol-dependent enzymes and chelating fungicides inhibit metal-dependent enzymes. The enzymes inhibited by captan in Table 2 are undoubtedly thiol-dependent. Hence, specificity of the toxicant depends upon the nature of the functional groups on the apoenzyme.

Although noncompetitive inhibitors are largely nonspecific, a slight change in their chemical structure may increase their specificity. According to BAKER (1967), less reactive inhibitors may require a second functional group of the enzyme juxtaposed to the first to bind to the enzyme. The necessity for a companion site on the enzyme delineates the types of enzymes attacked. Therefore, specificity can increase when the toxicant requires two or more sites for combining with the apoenzyme.

*Coenzyme Inactivation.* Captan and dichlone inhibit the conversion of acetate to citrate in *Neurospora sitophila* by inactivating Coenzyme A [OWENS and BLAAK, 1960 (1)]. In the presence of captan, CoA is kept in the oxidized form and, thus,

Table 2. *Enzymes inhibited by captan*

| Enzyme | Source | Reference |
|---|---|---|
| Aldolase | Yeast | MONTI and SISLER, 1962 |
| Glyceraldehyde-3-phosphate dehydrogenase | Yeast | MONTI and SISLER, 1962 |
| Glutamate dehydrogenase | *S. laxa* | BYRDE, MARTIN and NICHOLAS, 1956 |
| Hexokinase | Yeast | DUGGER, HUMPHREYS and CALHOUN, 1959 |
| Pyruvate decarboxylase | Yeast | HOCHSTEIN and COX, 1956 |

cannot form acetyl-CoA with acetate. Dichlone inactivates CoA by forming addition products with the coenzyme.

The inhibition of yeast pyruvate decarboxylase by captan presents an interesting case. HOCHSTEIN and COX (1956) found that thiamine pyrophosphate, the coenzyme, can reverse the inhibition and the mechanism of inhibition is considered to be competitive. However, since captan is a thiol reagent [LUKENS and SISLER, 1958 (1); OWENS and BLAAK, 1960 (2)], the fungicide may oxidize the coenzyme to thiamine disulfide. The active form of thiamine is the thiol form (BANHIDI, 1960). Additions of thiamine to the system may destroy remaining fungicide and supply the coenzyme to revive enzymic activity.

Fungitoxic chelating agents may rob metallic cofactors from enzyme systems (ZENTMYER, 1944). Evidence to support this mechanism is the inhibition of aconitase, an iron-dependent enzyme, by nabam, maneb, copper dithiocarbamate, and ziram (OWENS, 1963). Ferbam, the ferric salt, is the most stable dithiocarbamate and does not inhibit aconitase. The effective dithiocarbamates may exchange their metal for the ferric ion of aconitase.

*Substrate Alteration.* A toxicant may act by combining with an important substrate to make that substrate no longer usable by enzyme or organism. Methylol melamine, the wrinkle-proofing material for cotton fabrics, combines with the cellulose fibrils to prevent access of cellulase of the fibrils (BERARD, LEONARD and REEVES, 1961). The enzymic inhibitor protects cotton cloth from attack by mildew fungi.

The importance of the oxidation of glycolic acid to glyoxylic acid in sporulation of *Alternaria solani* prompted HORSFALL and LUKENS (1968) to examine a series of phenyl hydrazines and related compounds for antisporulant properties. Since hydrazines react with α-keto moieties to form hydroazones, HORSFALL and LUKENS reasoned that phenyl hydrazines would react with glyoxylic acid to make that compound unusable to the organism in sporulation. Most hydrazine-type compounds prevented sporulation and hydrazone-type compounds were inactive. Apparently, phenyl hydrazines act as antisporulants by reacting with glyoxylic acid, the substrate for an enzyme in the metabolism of sporulation.

## Conclusions

Fungicides may act on the synthesis or activity of enzymes. The mechanisms of action may proceed by one or more of several types. Fungicides can increase the synthesis of certain inducible enzymes and certain analogues of biochemicals may cause the synthesis of faulty enzymes. Antimetabolites produced by fungi may repress synthesis of catabolic and biosynthetic enzymes. Activity of endogenous enzymes may be stimulated by fungicides that act in place of cofactors in the enzymic system. Enzymic activity can be inhibited in one of several ways. Antimetabolites may act as false effectors of regulatory enzymes and inhibit activity of enzymes of a biosynthetic pathway. Other toxicants may attack functional groups of the apoenzyme inside or outside of the catalytic center. Fungicides may compete with substrate and coenzyme for binding sites in the reaction center of the enzyme. Certain fungicides react with cofactors to prevent them from functioning as coenzymes and a few toxicants may attack the substrate of enzymes to make the substrate useless to enzyme and fungus.

Chapter 8

# Structure-Activity Relationships

## Introduction

Most fungicides available today were discovered empirically. The information from these studies can be used to integrate structural characteristics of a fungicide with its toxic action. The characteristics so identified portray, in part, the nature of fungitoxic action.

Long ago, MEYER and OVERTON discovered, independently, with aliphatic acids, that the narcotic action of organic compounds is related to the partitioning of the compounds between oil and aqueous phases. The relation of fungicides to their partitioning in oil and water was noted by RICH and HORSFALL (1952), and they suggested that this physical property is a measure of the extent to which the toxicant penetrates the lipoid barrier in fungous membranes. The lipophilic portion of the molecule they called the lipophore and distinguished it from the reaction center — the toxophore. HANSCH *et al.* (1965) reemphasized the distinction between permeation and reactions of toxicants at the site of action. Oil:water partitioning was designated as a measure of permeation of a toxicant. They suggested also that the reactions at sites of action can be expressed by the electronic effects of the substituents on the reaction center of the molecule. However, LUKENS and HORSFALL (1967, 1968) have shown that both permeation and toxic reactions of fungicides are governed, in part, by both chemical properties.

In this discussion of chemical constitution and fungitoxicity, consideration will be given first to the fungitoxic moiety, second to substituents that promote permeation, and third to mathematical expressions of the two factors of fungitoxic action. For further information on structure-activity relationships of fungicides, the reader is referred to ALBERT (1960), HORSFALL (1956), SEXTON (1953) or SPENCER (1963).

## Fungitoxic Reaction Centers

A fungitoxic reaction center may be described as a molecular structure capable of combining with fungous components to affect the fungus in an adverse way. Evidently, the number of possible fungitoxic reaction centers is endless. The number reported is extremely high. The reaction center can encompass the entire molecule or it can be localized. More than one can be combined in a single molecule. To illustrate the diversity in structure of these compounds, a selection of fungitoxic compounds is given in Table 3. Since the action of most of the examples has had little investigation, the reaction center of many is not known.

Table 3. *A selection of fungitoxic componnds*

| Structure | Reference |
|---|---|
| Amines | |
| aliphatic | ECKERT *et al.*, 1961 |
| N-alkyl-1,3-propylene diamines | HUECK, ADEMA and WIEGMANN, 1966 |
| aromatic trichloroacetamides | SHOMOVA, RUDAVSKII and KHASKIN, 1965 |
| Arsenicals | |
| methyl arsines | SHIOYAMA, *et al.*, 1964 |
| phenarsines | GRANIN *et al.*, 1965 |
| phenylene and phenoxy arsines | ANDREEVA *et al.*, 1965 |
| Azo compounds | |
| arylazomethanes | ZSOLNAI, 1965 |
| *p*-hydroxy-phenylazoformamide | HAMS, COLLYER and HOUSLEY, 1965 |
| sodium *p*-dimethylamino benzene diazosulfonate | HILLS and LEACH, 1962 |
| Benzene | |
| chorinated nitrobenzenes | LAST, 1952 |
| diphenyls | HORSFALL, CHAPMAN and RICH, 1951 |
| diphenylethylenes | LUCKENBAUGH, 1964 |
| Bismuth | |
| trialkyl bismuth chloride | LEEBRICK, 1965 |
| Boron | |
| *o*-substituted phenyl borates | WECK and STERN, 1964 |
| Carbonates | |
| phenyl carbonates | PIANKA and POLTON, 1966 |
| phenylthio carbonates | PIANKA and POLTON, 1966 |
| trithio carbonates | HORSFALL and LUKENS, 1965 (1) |
| Cationic | |
| alkylammonium | MASON, BROWN and MINGA, 1951 |
| guanidine | BROWN and SISLER, 1960 |
| imidazoline | WELLMAN and MCCALLAN, 1946 |
| copper naphthenate | PTITSYNA and DURDINA, 1961 |
| copper oxychloride | HORSFALL, MARSH and MARTIN, 1937 |
| copper sulfate | |
| Bordeaux mixture | |
| mercury, salts, organo, and organic | |
| pyridinium salts | LOCICERO, FREAR and MILLER, 1948 |
| tin, organo | BENNETT, 1965; ZEDLER, 1966 |
| Dithiocarbamates | |
| dialkyl | TISDALE and FLENNER, 1942 |
| monoalkyl | |
| sodium methyl dithiocarbamate | HUGHES, 1960 |
| sodium ethylene-bis-dithio-carbamate | DIMOND, HEUBERGER and HORSFALL, 1943 |
| Mylone | VAN DER KERK, 1956 |
| Diheterocyclics | |
| tetrahydro-1,4-oxazines | KOENIG, POMMER and SANNE, 1965 |
| oxathiin | VON SCHMELLING and KULKA, 1966 |
| ethylene thiuram monosulfide | LUDWIG, THORN and MILLER, 1954 |
| mercaptothiazoles | SCHMITT, 1951 |
| thiadiazoles | THORN and LUDWIG, 1958 |

Table 3. Continued

| Structure | Reference |
|---|---|
| Ethers | |
| 3-phenoxy-1,2-epoxypropane | Bomar, Jedlinski and Hippe, 1966 |
| thioethers | Horsfall and Rich, 1953 |
| aryl-phenylpropyl-2,3-ene | Marcos, Municio and Vega, 1964 |
| Ethionine | Moje, Kendrick and Zentmyer, 1963 |
| Fatty acids | Hoffman, Schweitzer and Dalby, 1939 |
| Fatty acids and alcohols | Baechler, 1939 |
| Halo alkyl | |
| dichloropropane | |
| dichloropropene | |
| ethylene dibromide | |
| hexachloro butadiene | Deniskina *et al.*, 1966 |
| methyl bromide | |
| Halo alkyl thio | |
| trifluorodichloromethylthio | Kuehle, Klauke and Grewe, 1964 |
| tetrahalosulfenyl halide | Weil, Geering and Smith, 1964 |
| trichloromethylthio | Kittleson, 1953 |
| dithiadiazines | Sosnovsky, 1956 |
| Heterocyclic nitrogen | |
| ethylenethioureas | Rich and Horsfall, 1954 (2) |
| imides | |
| cycloheximide and isomers | Siegel, Sisler and Johnson, 1966 |
| *N*-imide-$SCCl_3$ | Kittleson, 1953 |
| itaconimides | Von Schmelling, 1962 |
| *N*-phenylmaleimides | Torgeson, Hensley and Lambrech, 1963 |
| imidazoline | Wellman and McCallan, 1946 |
| tetrahydropyrimidine | Radar, Monroe and Whetstone, 1952; Nickell, Gordon and Goenaga, 1961 |
| pyrazoles | McNew and Sundholm, 1949 |
| pyridine | |
| acridines | Horsfall and Rich, 1951 |
| 2-pyridinethione-*N*-oxide | Shaw *et al.*, 1950 |
| quinoline | Rigler and Greathouse, 1941 |
| 8-quinolinol | Meyer, 1932 |
| phosphoryl triazoles | Koopmans, 1960 |
| *s*-triazines | Schuldt and Wolf, 1956 |
| Heterocyclic oxygen | |
| furfural | Flor, 1927 |
| furfuron, nitro | Momoki *et al.*, 1964; Novikov *et al.*, 1963 |
| *m*-dioxolanes | Horsfall and Lukens, 1965 (1) |
| *m*-dioxanes | Horsfall and Lukens, 1965 (1) |
| methylenedioxybenzenes | Horsfall and Lukens, 1965 (1) |
| Hydroxamic acid | |
| arylthioalkylhydroxamic acids | Zayed, Mostafa and Farghaly, 1966 |
| arylhydroxamic acids | Buraczewski *et al.*, 1964 |
| Ketones | |
| halomethylarylketones | Lukes and Williamson, 1965 |
| β-aminoethyl ketones | Pellegrini, Bugiani and Tenerini, 1965 |
| α,β-unsaturated ketones | |
| quinones | McNew and Burchfield, 1951 |
| 2-methyl-3-thio-1,4-naphthoquinones | Wagner *et al.*, 1964 |

Table 3. Continued

| Structure | Reference |
|---|---|
| Nitrile | |
| aminohydroxystearo nitriles | HUECK, ADEMA and WIEGMANN, 1966 |
| tetrachloroisophthalonitrile | TURNER *et al.*, 1964 |
| Phenol | |
| nitrophenol | GRUENHAGEN, WOLF and DUNN, 1951 |
| nitrophenolic esters and ethers | YARWOOD, 1951 |
| | KIRBY, FRICK and GRATWICK, 1966 |
| | BYRDE, CLIFFORD and WOODCOCK, 1966 |
| naphthols | WOODCOCK and BYRDE, 1963 |
| resorcinol | WILCOXON and MCCALLAN, 1935 |
| Phosphorus | |
| bis(dimethylamido)pentachlorophenyl phosphates | SCHOOT, KOOPMANS and VAN DEN BOS, 1964 |
| dialkylthionothiol phosphoric acid esters | ANON, 1964 |
| phosphinamidothionateimidazol-1-yl | TOLKMITH *et al.*, 1967 |
| Propanediols | CHEYMOL *et al.*, 1964 |
| Silylacetylenes | MERKER, 1965 |
| Sulfonic acids- alkylthiosulfonates | WEIDNER and BLOCK, 1964 |
| Thiocyano- | |
| thiocyanates | WILCOXON and MCCALLAN, 1935 |
| isothiocyanates | WILCOXON and MCCALLAN, 1935 |
| | NEMEC *et al.*, 1964 |
| isothiocyanide | ANON, 1965 |
| Thiourea | |
| phenyl thiourea | SIJPESTEIJN and PLUIGERS, 1962 |
| ethylene thiourea | RICH and HORSFALL, 1954 (2) |

The diversity of structure of the fungitoxic compounds reemphasizes the lack of specific sites of action for most fungicides. The major restriction placed upon fungitoxicity is the migration of the toxicant to sites of action. However, the action of glyodin, cycloheximide, chloroneb, Botran and ethionine may be limited to a few specific sites.

## Metallic Ions

The fungitoxicity of metallic ions has been examined from various points of view. MCCALLAN and WILCOXON (1934) examined fungitoxicity of the elements in relation to their position in the periodic table. Generally, toxicity within a group increased with atomic weight. Silver and osmium were the most toxic elements. Fungitoxicity of copper, mercury, cerium, cadmium, lead, thallium and chromium varied with fungous species.

In addition to atomic weight, toxicity of metals to various organisms has been shown to be related to electronegativity of metallic ion, stability of metal chelate, and stability of the metal sulfide (HORSFALL, 1956; SOMERS, 1961; and SHAW, 1954, respectively). These relationships are compared in Table 4. The data indicate a relationship between chemical property and toxicity. The metals fall into three groups, within which the fungitoxicity varies. The metals are, in descending order of activity: Group I, Ag, Hg, and Cu; Group II, Cd, Cr, Ni, Pb, Co, and Zn; and Group III, Fe and Ca. Electronegativity can be considered an orbital as well as a nuclear property. The degree of electronegativity is a measure of the stability of metal bonds with cellular constituents. The influence of electronegativity on the stability of metal bonds is expressed in the stability of metal chelates and metal sulfides. These are the types of bonds involved in the fungitoxicity of metallic ions.

Apparently, the ease with which metals can enter covalent and coordinate bonds with biological constituents determines fungitoxicity of metallic ions. The mechanisms of transporting across fungous membranes and through the protoplast and the mechanism of reactions at the sites of action include, in part, these reactions of metallic ions.

Table 4. *Fungitoxicity and chemical properties of metallic ions. Order of Descending Activity*

| Fungitoxicity[a] | Electronegativity[b] | Chelate stability[c] | Insolubility of sulfide[d] |
|---|---|---|---|
| Ag | Cu | Hg | Hg |
| Hg | Hg | Cu | Ag |
| Cu | Ag | Ni | Cu |
| Cd | Co = Ni | Pb | Pb |
| Cr | Pb = Cr | Co = Zn | Cd |
| Ni | Zn | Cd | Zn |
| Pb | Mn | Fe | Ni |
| Co | Mg | Mn | Co |
| Zn | Ca | Mg | Mn |
| Fe | | | |
| Ca | | | |

[a] Data of Horsfall (1956) compiled from 21 references.
[b] Data of Haissinsky as reported by Somers (1961).
[c] Data of Melor and Maley (1948).
[d] Data of Shaw (1954).

## Sulfur

Elemental sulfur is an ancient fungicide. For the past two centuries, its fungitoxic properties have been exploited to control plant pathogens. However, in spite of intensive investigations by many researchers, it is not certain whether sulfur or a compound of sulfur permeates fungous cells and causes toxicity. Excellent reviews of studies on the toxic action of sulfur have been written by Horsfall (1956) and Martin (1964).

Compounds in which sulfur is oxidized or reduced have been invoked as the toxic species of the element. However, no one compound can explain entirely the responses of fungi to sulfur. Horsfall (1956), as well as Sempio (1932) before him, concluded that sulfur must be toxic by itself. Particles of sulfur have been observed in fungous cells that were treated with polysulfides. Because of its extremely hydrophobic nature, sulfur must permeate fungous cells by a nonaqueous mechanism. Since fungi sensitive to sulfur reduce the element to hydrogen sulfide, Horsfall (1956) suggested that sulfur competes with oxygen for protons and electrons in fungous respiration.

TWEEDY (1964) found that sulfur receives electrons between cytochromes b and c of this electron transport system of treated cells. OWENS (1960) suggested that the toxic entity is a $S_6$ molecular structure. The $S_6$ molecular form of sulfur, as a free radical, can react with cellular constituents. Sulfur and other fungicides capable of forming free radicals attack at least one enzyme in the conversion of acetate to citrate in conidia of *Neurospora sitophila* (OWENS, 1960).

## Dithiocarbamates

The dithiocarbamates are by far the most important group of organic fungicides for controlling plant diseases. The fungitoxicities of these compounds, investigated intensively, have been covered in a monograph by THORN and LUDWIG (1962) and reviewed by VAN DER KERK (1959), SPENCER (1963) and MOREHART and CROSSAN (1965). The family of compounds is divided into two groups on the basis of their chemical structure. The two types of dithiocarbamates, the dialkyl and the monoalkyl, differ in their biological actions (VAN DER KERK, 1959).

Tetramethylthiuram disulfide was the fungicide with which DIMOND *et al.* (1941) discovered the polymodal dosage-response curve. This peculiar curve has been found typical of fungitoxic metal binding agents, too. The sulfides and metal salts are more fungitoxic than esters of dithiocarbamic acid. Activity is greatest with the dimethyl derivatives and decreases with increase in length of the alkyl groups. Because of the decrease in induction of electronic effects by alkyl substitutions and esterification, the fungitoxic moiety was suggested to be as follows:

$$\begin{matrix} R \\ R \end{matrix}\!\!>\overset{\oplus}{N}=C\!\!<\!\!\begin{matrix} S^{\ominus} \\ S^{\ominus} \end{matrix}\!\!>\!M^{\oplus}$$

(VAN DER KERK and KLÖPPING, 1952). Permeation of lipoid barriers can occur by the nonionizable copper chelate and reactions can take place in the above canonical form.

How the dithiocarbamate permeates fungous cells is in dispute. Though thiuram disulfide and the metal dithiocarbamates may permeate fungous cells unchanged, THORN and RICHARDSON (1964) have found that fungous cells convert, extracellularly, considerable amounts of these materials to the dithiocarbamate ion. They suggest that the dithiocarbamate ion, regardless of its source, may permeate as the cupric chelate. Because the actions of the various forms of dithiocarbamate on Krebs cycle enzymes of fungous spores differ, OWENS and HAYES (1964) suggested that permeation is not done by a common species of dithiocarbamate. The significance in fungitoxicity of the extracellular fungous conversion of the dialkyldithiocarbamate fungicides to the anionic form has yet to be clarified.

The principal monoalkyldithiocarbamate fungicides are derivatives of the ethylene-bis series: nabam, zineb, maneb, ethylene thiuram mono- and disulfide alone and mixtures thereof. Pure nabam, maneb, and possibly zineb, are nontoxic without previous exposure to air (MOREHART and CROSSAN, 1965). Presumably, air degrades the compounds to ethylene mono- and disulfide, carbon disulfide and sulfides. Each of these products has been invoked as the toxic entity of the ethylene bis-dithiocarbamates.

The complexity of the chemistry of the ethylene bis-dithiocarbamates causes difficulty in studying their mechanism of action. Fungitoxic data on homologues suggest that a reduction in electronic inductive effects on the thiouride ion reduces biological activity. However, the reaction center may be the mobile *N*-hydrogen as well as the thiouride ion.

The mobile *N*-hydrogen makes possible the conversion of the monoalkyldithiocarbamates to isothiocyanates, which are fungitoxic compounds. Fumigant action of sodium methyldithiocarbamate as well as that of 3,5-dimethyl-1,3,4-tetrahydrothiazine-2-thione, a cyclic ester of dithiocarbamic acid, proceeds by way of an isothiocyanate derivative of the dithiocarbamate (Munnecke, Domsch and Eckert, 1962; van der Kerk, 1959). Likewise, the fungitoxic entity of the ethylene bis-dithiocarbamates has been suggested to be ethylene-bis-diisothiocyanate (Klöpping and van der Kerk, 1951). When exposed to air, residues of nabam decompose to ethylenethiuram monosulfide and its polymer, among other products (Ludwig, Thorn and Miller, 1954). The monosulfide can be converted to the diisothiocyanate with the aid of a catalyst in nonaqueous systems.

Although isothiocyanates are highly fungitoxic and may be gaseous toxicants of monoalkyldithiocarbamates in soil, the role of isothiocyanate derivatives in the fungitoxicity of maneb and zineb and possibly nabam is uncertain. *Rhizoctonia solani* responds differently to sodium methyl-dithiocarbamate and methylisothiocyanate (Wedding and Kendrick, 1959). Morehart and Crossan (1965) and Owens (1960) found that isothiocyanates attack fungous respiration at sites different from those attacked by maneb and nabam. The analogous behavior of maneb and ethylene-bis-thiuram disulfide on fungi and plant disease control caused Morehart and Crossan (1965) to conclude that the disulfide is the active form of the ethylene-bis-dithiocarbamates.

## Quinones

Quinones and their reduced phenols have been widely used in fungicide and disinfectant formulations. With proper substituents, toxic action can be limited to specific groups of organisms. With the quinones, the fungitoxic reaction center may be the ketone or its $\alpha$, $\beta$ unsaturated carbons. Substituents that increase reactivity about these centers also increase toxic action, but beyond a point where potency is lost, presumably, through reactions with noncritical components of the organisms. Substituents may affect efficacy, also, by altering permeation qualities of the toxicant through hydrophobic effects. In addition, substitution at the position *ortho* to the quinone may block access to the ketone and decrease activity of this center. McNew and Burchfield (1951) reported that fungitoxicity decreases with change in quinone nucleus accordingly: naphthoquinone > phenanthraquinone > benzoquinone > anthraquinone. Activity increases with halogenation and with the halogen I < Br < Cl. Since halogen substitution increases fungitoxicity and alkyl substitution has the opposite effect, fungitoxicity requires mobile substituents to the unsaturated carbon atoms $\alpha$, $\beta$ to the ketone. Reactions at these sites are involved in the mechanisms of action.

Fungitoxicity of phenol is increased with proper substitution to the benzene ring. Chlorination at the 2,4,6-positions causes maximum antisporulant action of chlorophenol (Lukens and Horsfall, 1968). Maximum wood preservation occurs with pentachlorophenol. Systemic action against bean mildew by phenol requires an *ortho* or *para*-nitro substituent, an alkyl substituent and a halogen substituent. One nitro group is as effective as two. With alkyl substituent, systemic activity increases in the order of substitution to the *ortho*-position: phenyl $<$ cyclohexyl $<$ 1′-methylheptyl. *Ortho*-chloro-substitution is more effective than *ortho*-bromo-substitution. The compound found most active by El-Zayat *et al.* (1968) was the multibranched 4-(1′,1′,3′,3′-tetramethyl)butyl 2-chloro-6-nitrophenol. With the alkyldinitrophenols, the 4-alkyl-2,6-dinitro analogues are better mildewcides than the 2-alkyl-4,6-dinitrophenols (Byrde, Clifford and Woodcock, 1966). The most active compound of the 4-alkyl series is the 1-′ethylhexyl derivative (Byrde, Clifford and Woodcock, 1966; Kirby, Frick and Gratwick, 1966) and of the 6-alkyl series, 1′-methylheptyl, the free phenol of Karathane (2,4-dinitro-6-(1′-methyl)heptyl phenyl crotonate).

Presumably, Karathane is toxic by virtue of the fungus hydrolyzing the compound to the free phenol. Kirby and Frick (1958) presented evidence to show that the free phenol is the most toxic derivative of Karathane. Capability of fungi to hydrolyze phenyl ethers is supported by the work of Hock (1968), who has isolated 2,5-dichloro-4-methoxyphenol from mycelium of *Rhizoctonia solani* which was treated with chloroneb (1,4-dichloro-2,5-dimethoxybenzene).

## Heterocyclic Fungicides

Since many metabolites necessary for growth of fungi are heterocyclic compounds, one would expect that fungicides containing the heterocycle should act as antimetabolites. Yet, glyodin is the only heterocyclic fungicide reported to act competitively. West and Wolf (1955) found it competes with guanine and xanthine. How glyodin competes with the purine and pyrimidine bases in protein and nucleic acid synthesis is not clear (Kerridge, 1958). The effect of substituents at positions 1 and 2 of the imidazoline ring on fungitoxicity can be explained in terms of hydrophobic effects (Rich and Horsfall, 1952). When the imidazoline ring is quaternized, toxicity follows surfactant activity of 2-alkyl homologues (Shepard and Shonle, 1947).

Tolkmith *et al.* (1967) have described some fungitoxic derivatives of imidazole. Fungitoxicity of imidazole substituted on the imino-nitrogen atom depends upon the electron-attacking properties of the substituent and necessitates that the atom attached to the nitrogen atom have a tetrahedral geometry. An imino-nitrogen atom is required for toxicity.

Apparently, activity of cycloheximide requires a structure that enables the glutarimide to make a three point attachment to substrate (Siegel, Sisler and Johnson, 1966). The groups of attachment are an unsubstituted imide group, a free hydroxyl group and a ketone moiety. In addition, the cyclic ketone moiety requires a proper orientation to the remainder of the molecule.

The toxic moiety of the cyclic imide fungicides — captan, folpet, and difolatan — is open to question. Originally, Kittleson (1953) suggested that the toxic portion of captan was the $-N-SCCl_3$ group. Horsfall (1956) countered with the cyclic imide portion since fungitoxicity of captan mimicked that of cycloheximide. The $-SCCl_3$ group has been shown, since, to confer toxicity to numerous $R-SCCl_3$ compounds (Lukens, 1966). R-groups forming toxic $R-SCCl_3$ compounds include cyclic imides, sulfonamides, sulfinates, and alkanes, some of which contain no nitrogen or imide.

Fungitoxic properties are tied in some way to the reaction of these compounds with thiols (LUKENS, RICH, and HORSFALL, 1965).

However, the imide theory has regained some importance. Cyclic imides of the phthalimide type exhibit fungitoxic properties provided their hydrophobic and electronic properties are adjusted correctly (LUKENS and HORSFALL, 1962, 1967). LIEN (1969) showed that toxicity and hydrophobic bonding of both imides and their *N*–$SCCl_3$ compounds can be included in the same parabolic equation. The equation, derived from the method of least squares, is significant at the 1% level. Apparently, the imide and –$SCCl_3$ group are potentially fungitoxic. To distinguish which group plays the greater role in the action of captan and its homologues is a tedious task.

VON SCHMELING (1962) reported fungitoxicity of itaconimide. Oxidation of the heterocycle to maleimide reduces fungitoxicity. The most toxic itaconimide was the *N*–(3-nitrophenyl) analogue. *Para* or *ortho* nitration of the *N*-phenyl substituent drastically curtailed activity, while chlorination of the *N*-phenyl group had minor influences.

The systemically active oxathiin fungicides require an aryl amide substituent at the 5-position (VON SCHMELING and KULKA, 1966). Alkyl amides and esters of carboxylic acid are inactive. Highly active analogues against basidiomycetous fungi were the 2,3-dihydro-5-carboxanilido-6-methyl-1,4-oxathiin and its sulfone analogue (EDGINGTON, WALTON and MILLER, 1966). However, a phenyl group substituted to the benzene ring *ortho* to the amido group is required for high activity against ascomycetous fungi (EDGINGTON and BARRON, 1967).

## Miscellaneous Compounds

Dodine (dodecylguanidine acetate) is the most fungitoxic of the alkyl-guanidine-acetate series (BROWN and SISLER, 1960). Presumably, the alkyl group influences activity through its hydrophobic effects on the molecule. Fungitoxicity may depend upon reactions of the imino group because activity increases with the pH of the medium.

Inhibition of mycelial growth by a series of nitroanilines is maximum with the 2-chloro-6-bromo derivative (CLARK and HANS, 1961). Activity decreases with replacement of halo-substituents dichloro > dibromo > diiodo. Activity falls off, with both amino hydrogen atoms replaced by alkyl groups, with both *ortho* positions of aniline occupied with nitro- or iodo-substituents, and with the amino group replaced by methoxy-, acetic acid or hydroxyl groups. Apparently, the optimum hydrophobic bonding falls within the series and the halogens affect toxicity through their hydrophobic properties. The curtailment of activity by removal of the mobile hydrogen atoms of the arylamine or blocking access to these atoms through steric hindrance suggests that the arylamine is an essential part of the toxophore of nitro-anilines. The compound 2,6-dichloro-4-nitroaniline has been developed commercially under the name Botran.

Chlorinated nitrobenzenes are toxic to a limited number of phytopathogenic fungi common in soil. Adding chlorine atoms to the molecule increases activity (ECKERT, 1962). With the tetrachloro analogue, the 2,3,5,6-tetrachloro isomer is far more effective than other isomers (BROOK, 1952). Presumably, both of these structural requirements promote activity by hastening the migration of toxicant to sites of action, the first by increasing hydrophobic bonding to permeate the membranes of the fungous cell and the second by increasing volatility to make contact with fungi in the soil. SPENCER (1963) suggested that activity may be due to reactive chlorines caused by the electron-withdrawing property of the nitro group. The nitrobenzenes may be toxic by reacting indiscriminately with cellular constituents (ECKERT, 1962).

## Permeation Centers

Certain chemical structures increase toxicity, presumably by hastening the movement of fungitoxic compounds across cell membranes and through the protoplast to sites of action. HORSFALL (1956) emphasized the importance of lipophilic substituents to the activity of a fungitoxic nucleus. Earlier, RICH and HORSFALL (1952) used the analogy of the shape of a charge in a missile which permits its penetration of an armored plate to describe the permeation reaction center necessary for a compound to pass through cellular membranes. Such a center was termed a "shaped-charge".

The most common "shaped-charge" since the works of MEYER and OVERTON is the aliphatic chain. By lengthening the chain of aliphatic substituents, activity of numerous fungitoxic moieties has been increased. Fig. 7 illustrates the effect of chain length on the toxicities of 2-substituted imidazolines. The length of the alkyl chain has been equated to the toxicity, the enzyme inhibition, and movement within xylem vessels of higher plants of alkyl quaternary ammonium compounds (ANDERSON and REYNOLDS, 1965 and EDGINGTON and DIMOND, 1964). Indivdual methylene groups can be added to several sites of the parent compound to increase fungitoxicity of *m*-dioxanes [HORSFALL and LUKENS, 1965 (1)].

Halogen atoms have been shown to bring about the same effect as aliphatic groups. Fungitoxicity of phenols and nitrobenzenes increases with chlorination. Also, the inhibition of sporulation of *Alternaria solani* increases with chlorination of the phenoxy-acetic acid antisporulants (LUKENS and HORSFALL, 1968).

RICH and HORSFALL (1952) suggested that a measure of the shaped-charge is the change in the oil:water partition coefficient of the compound by the substituent. ANDERSON and REYNOLDS (1965) considered the energy of the C–C bond as representative of the free energy related to the hydrophobic effect, which provided a rational basis for plotting activity against carbon number or molecular weight. HANSCH and FUJITA (1964) introduced the substituent constant, $\pi$. $\pi$, a logarithmic function of oil:water partitioning, is defined as:

$$\pi = \log \frac{\text{Ps}}{\text{Ph}}.$$

Ps and Ph are the oil:water partition coefficients of the substituted compound and that of the parent compound, respectively. HANSCH and FUJITA (1964) considered $\pi$ to $\Delta$ change in free energy of an analogue relative to that of the parent compound in moving the derivative from one phase to another. Although $\pi$ is derived from equilibrium data, it is used in view of a flux process in serial partitioning of compounds through several hydro- and lipo-phases. The rate of flow, then, is proportional to the hydrophobic characteristic of the compound when that property is limiting.

There is a close agreement between $\pi$ values of substituents as determined from different base compounds (FUJITA, IWASA and HANSCH, 1964). For instance, methylation of the aromatic nucleus at the 3-position of phenoxyacetic acid, phenylacetic

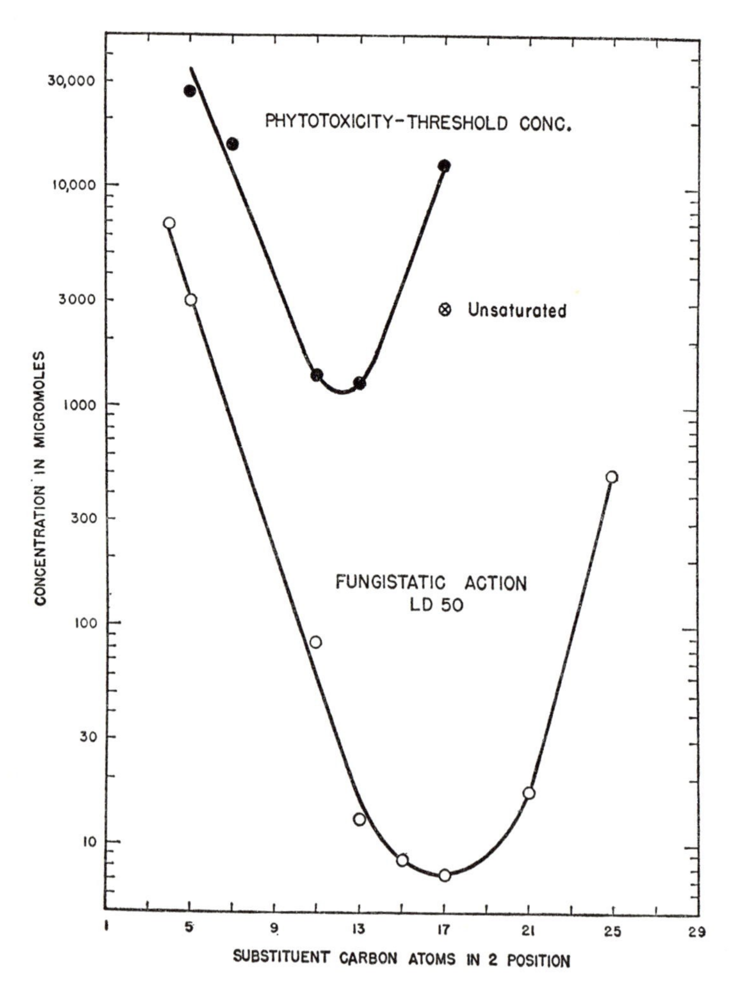


Fig. 7. Relationship between fungistatic action and phytotoxicity for a series of 1-hydroxyethylglyoxalidines. Note maximum spread between two curves at 17 substituent carbon atoms in 2-position. [From R. H. WELLMAN and S. E. A. MCCALLAN: Glyoxalidine derivatives as foliage fungicides. I. Laboratory studies. Contribs. Boyce Thompson Inst. **14**, 151—160 (1946)]

acid, benzoic acid, benzyl alcohol, phenol, aniline, nitrobenzene, and benzene gives $\pi$ values of 0.51, 0.49, 0.52, 0.50, 0.56, 0.50, 0.57, and 0.56 respectively. The $\Delta$ increase in the $\pi$ for addition of methylene fragments to 3-methylphenoxyacetic acid is 0.46. The reported $\pi$ values for 3-alkyls are: methyl, 0.51; ethyl, 0.97; *n*-propyl, 1.43; and *n*-butyl, 1.90.

Substituents having plus $\pi$ constants increase hydrophobic bonding and those with minus values decrease hydrophobic bonding. Components having plus $\pi$ values include fluoro, bromo, chloro, iodo, alkyl, aryl, methylmercapto, trihalomethylmercapto, and trihalomethoxy substituents. Substituents with minus $\pi$ constants are cyano, carboxylic, and hydroxyl. Values of $\pi$ are plus and minus depending upon the position of substitution on the aromatic nucleus for nitro, methoxy, and methoxyalkyl substituents. The lowering of $\pi$ ($-1.26$) with 3-methylsulfone can be overcome by

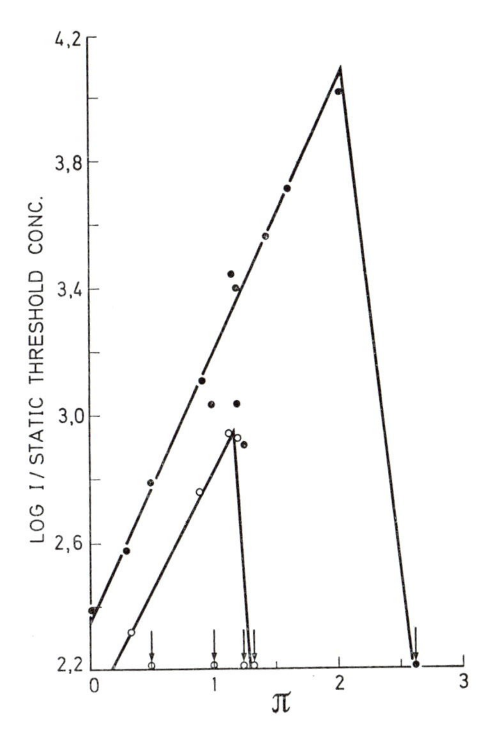


Fig. 8. Effect of hydrophobic bonding ($\pi$) on the toxicity of 5,7-substituted 8-methoxyquinoline to *Trichophyton mentagrophytes* (●—●) and to *Aspergillus aryzae* (○—○). Threshold values in moles per liter are taken from Gershon and Parmegiani (1968). The $\pi$ values of 8-methoxyquinolines are: no substitution, 0; 5-fluoro-, 0.31; 5-chloro-, 0.93; 5-bromo-, 1.13; 5-iodo-, 1.45; 5-nitro, 0.50; 5,7-dichloro-, 1.62; 5,7-dibromo-, 2.02; 5,7-diiodo, 2.64; 5-chloro-7-nitro-, 1.26; 5-chloro-7-fluoro-, 1.18; 5-fluoro-7-chloro-, 1.00; 5-fluoro-7-bromo-, 1.20. The $\pi$ values are taken from substituted phenol (Fujita, Iwasa and Hansch, 1964)

fluorination of the methyl groups. Value of $\pi$ for trifluoromethylsulfone was 0.93. One can see that, with $\pi$ analysis, a substituent will either increase or decrease the hydrophobic characteristic of the analogue. Moreover, the relative hydrophobic characteristic of an analogue equals the sum of $\pi$ values of the individual substituents. Since $\pi$ values of a substituent vary little between nuclei, they can be used with closely related nuclei to rationalize structure-activity relationships of apparently unrelated and puzzling structures.

To illustrate the usefulness of $\pi$ analysis, the data of GERSHON and PARMEGIANI (1968) on the growth inhibition of *Trichophyton mentagrophytes* and *Aspergillus oryzae* by 5,7-substituted 8-methoxyquinoline are converted to a suitable form and plotted against $\pi$ values of substituents in Fig. 8. The $\pi$ values are those of substituted phenol reported by FUJITA, IWASA and HANSCH (1964). Since quinolinol is a derivative of phenol, these $\pi$ values can suffice for a first approximation.

As hydrophobic bonding is increased, toxicity to both fungi increases, reaches a maximum, and sharply falls off. The fungous responses differ in the optimum for hydrophobic bonding and differ in potency at that optimum. These two characteristics can be a basis for selective action.

The toxic actions of 12 of the 13 analogues of methoxyoxine to *T. mentagrophytes* can be accounted for by the hydrophobic characteristics of the compounds. The exception, 5-chloro-7-nitro-8-methoxyoxine, has less activity than its hydrophobic bonding would suggest. This reduction cannot be explained by the electronic effects of the nitro group on the phenolic reaction center because the 5-nitro analogue does not deviate from the curve. Presumably, the nitro group *ortho* to the methoxy group reduces activity through steric hindrance.

With *A. oryzae* hydrophobic bonding of half the compounds exceeds the optimum for toxicity and they are nontoxic. Some compounds with hydrophobic bonding in the effective range are nontoxic, too. Presumably with these nontoxic compounds, an excessive withdrawl of electrons from the methoxygroup restricts toxicity to *A. oryzae*. The 5-nitro analogue is in this nontoxic group.

The paraboloid curves of fungous responses in Fig. 8 suggest the quadratic function

$$\log \text{Toxicity} = K\pi^2 + K'\pi + C,$$

where $K$ and $K'$ are constants of the two functions of $\pi$ and $C$ is a constant containing factors outside of hydrophobic bonding. The values of $K$ and $K'$, regression coefficients of the data to form an equation, are peculiar to fungous species. However, since the curves are not symmetrical on either side of the optimum value of $x$, the $K$- values must vary somewhat with $\pi$. If one restricts his study to compounds having $\pi$ values below the optimum for fungitoxicity where the equation is linear and the $K$ value is constant, the equation becomes:

$$\log \text{Toxicity} = K\pi + C$$

which is common to many examples of structure-activity data.

## Hydrophobic, Electronic, and Fungitoxic Relations

Since the hydrophobic bonding characteristic affects predominantly the migration of toxicant from the outside to sites of action within fungous cells, HANSCH and FUJITA (1964) proposed that $\pi$ represents essentially a permeation process in biological action. They suggest that reactions at sites of action can be represented by the electronic effects of substituents on the reaction center and used HAMMETT's substituent constant $\sigma$ to represent activity of the compound at the site of action. The overall equation for rationalizing structure-activity data of analogues relative to the unsubstituted compound becomes:

$$\log \text{Toxicity} = K'\pi^2 + K''\pi + K'''\sigma + K'''' + C$$

Since the few molecules of toxicant that reach the reaction sites have sufficient electronic induction to react with cellular constituents, $\sigma$ is not limiting activity and can be ignored in many situations.

As evidenced by recent work, excess of the electronic influence of the substituent can restrict permeation of the compound. Also, the inhibitors may bind hydrophobically to enzymes. With nonaromatic analogues of phthalimide, the weak fungitoxicity toward spore germination of *Stemphylium sarcinaeforme* is negative and linear to the *pKa* values of the imides (LUKENS and HORSFALL, 1967). Since the compounds permeate fungous cells in the undissociated form, an increase in electronic effects shifts the equilibrium away from the permeating species. Likewise, antimildew activity of captan and its homologues is negative and linear to the degree of reactivity of the compounds with thiols (LUKENS and HORSFALL, 1967). The thiol reaction has been considered a mechanism of detoxication and, thus, poses as a block to permeation of the fungous cells by the toxicants. The reactivity of the N–S center of captan, like that of the imide center of phthalimide, is affected by the electronic influence of the changes in the molecule.

Permeation and reactions at sites of fungitoxicity are distinctive biological actions and can, like fungitoxicity, be equated to functions of hydrophobic and electronic effects of substituents. In the linear function then:

$$\log \text{Permeation} = K\pi - K'\sigma + C$$

and

$$\log \text{Reaction} = K''\pi + K'''\sigma + C'$$

Since

$$\log \text{Fungitoxicity} = \log \text{Permeation} + \log \text{Reaction},$$

then

$$\log \text{Fungitoxicity} = (K\pi - K'\sigma + C) + (K''\pi + K'''\sigma + C')$$

When the values of all $K$ approach unity, the electronic effects cancel out and the equation becomes essentially

$$\log \text{Fungitoxicity} = 2\,\pi,$$

which is a return to the linear relationship of fungitoxicity to hydrophobic bonding. However, the hydrophobic bonding property reflected in structure-activity data affects both penetration to and reactions at sites of action.

## Conclusions

The literature abounds in reports of structure-activity data for fungitoxic compounds. An empirical approach to the data identifies fungitoxic moieties. When fungitoxic targets are defined more clearly, the search for fungitoxic moieties will be done on a rational basis. However, when toxicity is equated to the hydrophobic and electronic characteristics of substituents, the apparent chaotic interrelationships of the majority of compounds are sorted into rational order and the few exceptional analogues are isolated. Analyzing fungitoxic data with the hydrophobic and electronic substituent constants is useful in discerning properties of permeation and reactions at sites of action.

Chapter 9

# Action of Fungus on Fungicide

## Introduction

When a toxicant or its precursor enters the fungous cell, it may react in any of four ways: (1) it may inhibit the fungus directly; (2) it may accumulate in the fungous cell and be stored; (3) the precursor may be converted to a toxicant through metabolic conversion; or (4) the toxicant may be inactivated by the fungus. In all of these cases, a portion of the molecular structure of the toxicant or its precursor may remain unaltered in the cell or the material may be changed, either reversibly or irreversibly.

Normally, a toxicant is consumed when it inhibits a fungus directly. Most chemical reactants, *e.g.*, captan, dichlone, and dyrene, are irreversibly changed when they cause inhibition and cannot be recovered from treated cells (RICHMOND and SOMERS, 1962; OWENS and NOVOTNY, 1958; BURCHFIELD and STORRS, 1957). In the case of a few toxicants, the molecular structure remains intact and a portion which entered the cell can be recovered. Cycloheximide, Botran, and heavy metals have been leached from fungous cells that were inhibited by these toxicants (COURSEN and SISLER, 1960; WEBER and OGAWA, 1965; HORSFALL, 1956, respectively). When dithiocarbamates inhibit cells, they are converted to addition-products with cellular components and can be reconstituted by treating cells with thiols (OWENS and HAYES, 1964).

Mechanisms by which fungi store a toxicant, activate a precursor, and detoxify a toxicant are described below.

## Storing the Toxicant

When fungous cells take up a fungicide, they may store and sequester large amounts of the chemical. Storage implies that the toxicant is held in reserve for future use. The fungicide is stored by lowering of its activity coefficient in the aqueous phase. Activity coefficients denote that portion of a concentration of a chemical available for reaction. When the activity coefficient is low, only a minute amount of the fungicide within the cell is available for toxic reactions. Activity coefficients can be reduced by physical and chemical means.

### Physical Mechanisms

*Partitioning into Nonaqueous Systems.* Since most metabolic functions proceed in an aqueous environment, fungitoxic reactions are assumed to proceed likewise. Fungicides, being hydrophobic, are not very soluble in water. Their activity coefficients are high because of the low affinity of water for them (HIGUCHI, 1960). However, a large amount of fungicide will partition into nonaqueous systems of the cell where the toxicant is held firmly and its activity coefficient is low (HIGUCHI, 1960). Most of the glyodin taken up by conidia of *Neurospora sitophila* is found associated with lipids of the particulate fraction on cell disintegration (MILLER and RICHTER, 1960). In the lipoid phase, the activity coefficient of glyodin is reduced and reactions of the toxicant with the cell are discouraged.

Toxicants can be excluded from the aqueous phase by the polymeric compounds of the cell. Resting ascospores of *Neurospora* can store more than a lethal dose of methylene blue, phenyl mercury, or silver in their cell walls (SUSSMAN, 1963). Apparently, conidia of *Neurospora* detoxify dodine by binding it to their cell walls (SOMERS and PRING, 1966).

*Osmotic Effects.* The activity coefficient of a solute in water is inversely proportional to the concentration of other solutes. In dilute solutions of solute normality 0.01, activity coefficients are essentially unity. Most growing media for fungi are in this category. However, as the concentration of solutes increases beyond 0.01 N, the activity coefficient becomes less than unity. In physiology, the effect of solute strength has been examined from the view-point of water relations and the term osmotic pressure is employed to describe that force necessary to prevent the movement of water from dilute to concentrated phases. Thus, a mole of nonionized solute in one liter of water at 0 °C has an osmotic pressure of 22.4 atmospheres if separated from pure water by a semipermeable membrane. The average osmotic pressure of growth media is about 0.3 atmospheres. Growth rates of fungi approach zero when the osmotic pressure of solution reaches 30 to 60 atmospheres in syrups and brines.

The osmotic pressure of fungous cells varies with species and with the environment. COCHRANE (1958) reported that the osmotic pressure of hyphae of pathogenic fungi varies between 15 and 40 atmospheres and that of haustoria, between 19 and 22 atmospheres. When grown on concentrated media, *Aspergillus glaucus* var. *tonophilus* can develop an osmotic pressure of 250 atmospheres.

In the field, when moisture is deficient, activity coefficient of fungicides in the aqueous phase of fungous cells becomes low. Hence the fungicide is stored until moisture becomes available. Thus, the success of a fungicide in the field may depend, in part, upon its uptake through nonaqueous systems and its ability to produce toxic reactions in spite of a reduced activity coefficient in the aqueous phase under conditions of water stress. The toxicity to spores or mycelium in agar or liquid media does not measure the performance of a fungicide under water stress.

## Chemical Mechanisms

*Conversion to the Undissociated Form.* A large reserve supply of chlorine in swimming pool water can be maintained by a mildly alkaline pH when exposure time is not critical to sanitation. However, when short exposures are required for disinfection, the same pH can nullify the toxicity of chlorine (SHANNON, CLARK and REINHOLD, 1965). In many reactions, the dissociated form of a compound is the one that participates. The pH within the fungous cell determines the extent of dissociation of the toxicant within the cell. When the toxicant is stored in the undissociated form, it is unreactive.

Heavy metals hydroxylated at alkaline pHs are precipitated from the aqueous phase as the pH of cellular fluid rises. Since fungitoxicity is due to the metallic ion, hydroxides can be considered as storage of metal in the undissociated form by fungous cells.

Fungitoxic chelates may be stored by fungi in the form of the full chelated metal. Excess of natural chelator stabilizes the undissociated form of copper 8-hydroxyquinolate (ZENTMYER, RICH and HORSFALL, 1960).

*Conversion to Addition-Products.* The thiol-containing fungicides sodium dimethyldithiocarbamate and sodium pyridinethione-*N*-oxide are condensed to the terminal carbon of $\alpha$-amino and $\alpha$-keto butyric acids to form addition-products of the toxicants by fungi (SIJPESTEIJN and VAN DER KERK, 1965). Since fungitoxicity of these addition-products is considered to proceed by reconversion of the compound by the organism to the original toxicant, the formation of addition-products by the fungus represents a mechanism of storage. If the addition-product is nontoxic, however, its formation by fungi would represent a mechanism of detoxification.

According to SIJPESTEIJN *et al.* (1963) dimethyldithiocarbamic acid and pyridinethione-*N*-oxide compete with cysteine in the synthesis of methionine. In the normal process, the thiol of cysteine reacts with the hydroxyl site on homoserine to form cystathione (HORWITZ, 1947). Hence, fungi synthesize $\gamma$-(dimethylthiocarbamoylthio)-$\alpha$-amino butyric acid from dimethyldithiocarbamate. The $\alpha$-amino group is then removed to form the keto analogue. In green plants, methionine synthesis proceeds by condensing homocysteine with serine and plants convert the dithiocarbamate to the alanine addition-product.

## Producing the Toxicant

Fungi produce a toxic agent by degrading a precursor, or by synthesizing abnormal enzymes from an antimetabolite. Both processes may be considered lethal synthesis (MARKHAM, 1958).

### Degrading the Precursor

Fungi can reverse pathways of storage to release the toxic agent from reservoirs in storage. Oil:water partitioning, osmotic effects, and dissociation are equilibrial phenomena. A shift in the equilibrium can release the toxicant from storage. A change in cellular pH can increase dissociation of toxicants in the aqueous phase. An increase in dissociation in the aqueous phase can shift the partitioning of solutes to the aqueous phase from the lipoid phase. In addition to pH the dissociation of the full copper chelate of 8-hydroxyquinoline is dependent upon the availability of amino acids and other chelating agents in the cell (ZENTMYER, RICH and HORSFALL, 1960). Amino and keto acid excretions from fungous spores solubilize copper from dried residues of Bordeaux mixture (MCCALLAN and WILCOXON, 1936). When water stress of fungi is relieved, toxic reactions inhibited by high osmotic conditions can then proceed.

A common form of fungitoxic precursor is addition-products and esters or ethers from which the fungus liberates the toxicant. The $\beta$-glucosides and amino acid addition-products of dimethyldithiocarbamate can be hydrolyzed by fungi, thus liberating the toxicant again (SIJPESTEIJN *et al.*, 1963). When applied to diseased plants, esters of dithiocarbamates were selectively toxic to fungi that hydrolyzed the esters, thus freeing the toxic dithiocarbamate ion (VAN DER KERK, 1956).

Fungitoxic precursors administered to fungi are bioconverted to toxic agents through oxidation, de-esterification and hydrolysis. Phenols become fungitoxic on oxidation to quinones [OWENS, 1953 (2)]. Fungitoxicity of captan and other R–$SCCl_3$ compounds may arise, in part, from thiophosgene, which is formed *in vivo* on the reduction of the trichloromethylthio group by cellular thiols [LUKENS and SISLER, 1958 (1)].

Fungi may hydrolyze ether and ester analogues of dichlone to toxic products (BYRDE and WOODCOCK, 1953). The rates of alkaline hydrolysis of the compounds to yield 2,3-dichloro-1,4-naphthaphenol proceed in the same order as that of fungitoxicity of the compounds. Hydrolytic rates and toxicity to *Sclerotinia laxa*, *Botrytis fabae*, and *Cladosporium fulvum* increase with substitution to the phenolic group in the order: methyl = benzoyl < butyl < propionyl < acetyl. In addition, fungitoxicity of a mixture of the diacetyl analogue and an esterase is indistinguishable from that of dichlone (the quinone form) alone. Presumably, the ester is hydrolyzed to the hydroquinone which, in turn, is oxidized to the quinone *in vivo*. HORSFALL (1956) suggested that the oxidative process is by way of polyphenol oxidase. In like manner, phosphate esters of 1,4-dihydroxynaphthalene are acted upon by alkaline phosphatase in bacteria as the first step in forming 1,4-naphthoquinone (MARRIAN, FRIEDMAN and WARD, 1953).

Phenolic esters, on hydrolysis, form toxic phenols. Under suitable conditions, Karathane is degraded to 2,4-dinitro-6(1-methyl-*n*-heptyl) phenol and crotonic acid. KIRBY and FRICK (1958) have shown that control of the powdery mildew disease by Karathane is due to free 2,4-dinitro-6-(1-methyl-*n*-heptyl)phenol. Ammonium crotonate had little anti-mildew activity. The hydrolysis of Karathane to its phenolic and acidic components may occur spontaneously when the spray residues dry on plant surfaces (RICH and HORSFALL, 1949). However, the actions of nitrophenolic esters against sporulation in *Alternaria solani* suggest that the free phenol is released *in vivo*. HORSFALL and LUKENS [1966 (2)] reported no activity with Karathane but high activity with its acetate and pentenoic acid analogues and with 2,4-dinitro-6-(1-methyl-*n*-heptyl)phenol.

Apparently, esters and oxime of cycloheximide are toxic by virtue of their bioconversion to cycloheximide (SISLER, SIEGEL and RAGSDALE, 1967). The acetate ester is as toxic as the parent compound to tomato plants and nontoxic to yeast cells. Yeast may not have an esterase to release free cycloheximide. When the acetate ester is acted upon by esterase, the resulting product is as toxic as cycloheximide to *Saccharomyces pastorianus*. With other fungi, toxicity of esters is $^1/_{32}$ to $^1/_{64}$ that of cycloheximide. The release of cycloheximide from the oxime and semicarbazone derivatives is pH-dependent, and it would appear that breakdown *in vivo* of these derivatives is by chemical means.

With particular cases in chemotherapy of plant diseases, the host may convert the chemotherapeutant to the toxicant. Phenylthiourea, effective against scab of cucumber, is converted to an oxidized component by the host, presumably by polyphenol oxidase (VAN DER KERK, 1963).

### Synthesizing Faulty Enzymes

Thiouracil, fluorouracil and other analogues of pyrimidine may be toxic by virtue of their incorporation into nucleic acids and subsequent enzyme systems that may malfunction (BROCKMAN and ANDERSON, 1963). The mistake in identity of a synthetic compound for a metabolite on synthesis of malfunctioning catalysts can be called faulty synthesis. Fungitoxicity of 6-azauracil is an example of the faulty synthesis. DEKKER and OORT (1964) found that the pyrimidine is converted in cucumber plants to 6-azauridine. Since both compounds prevent powdery mildew disease, the toxic agent may be the uridine monophosphate. The toxic action is reversed by uracil precursors. They suggested that azauracil is converted to a riboside-containing metabolite which inhibits orotidylic decarboxylase in the synthesis of pyrimidines.

8-Hydroxyquinoline may participate in mechanisms of faulty synthesis. The chelator has been shown to cause the copper to be inserted into organic configurations where other metals normally function. LOWE and PHILLIPS (1962) found that in the presence of the chelator, copper is inserted into dimethyl porphyrin *in vitro*. However, copper porphyrin could not be found in bacterial cells treated with the chelator (JONES, 1963).

Faulty synthesis is a recent concept of fungitoxic mechanisms. Surely, with advances in fungous metabolism and especially those in microbial deterioration, faulty synthesis will be exploited in developing new agents for fungous control.

## Detoxifying the Toxicant

Fungitoxic action is the result of permeation of toxicant and reactions at sites of action. If either of these properties is destroyed, detoxication will occur. On adapting to and resisting the action of toxicants, fungi may (1) block binding and target sites, (2) continue growing in an altered manner in spite of toxic action, or (3) detoxify the toxicants. The first is an example of direct reaction to toxicity (active resistance). The second involves endurance (passive resistance). In the third, the toxicant is decomposed by the fungus (direct attack).

Hence, detoxication conveys resistance and can be a mechanism of fungous adaptation to toxicants. A fungicide is detoxified when its paths to sites of action are blocked. The blocking mechanism may be nonreversible storage or degradation of the fungitoxic molecule.

### Nonreversible Storage Mechanisms

The mechanisms of storage of fungicides, depending upon the circumstance, are considered to operate both ways: movement of fungicide into storage and movement of fungicide out of storage. However, where the fungus fails to mobilize the fungicide from storage, the storage mechanisms are, in essence, modes of detoxication.

The conversion of toxic metal ions to metal-chelate complexes by a natural chelating agent can lead to detoxication provided a high concentration of the natural chelator is maintained by the fungus. Apparently, the failure to control damping-off disease by thiram lies in the excess of natural chelating compounds in exudates of seed and roots, which interfere with the copper chelated toxicant (RICHARDSON, 1966).

In higher plants, the dithiocarbamate fungicides are converted to nontoxic addition-products. The plant, apparently unable to release the toxic moiety from the addition-product, detoxifies the dithiocarbamates by this storage mechanism. Fungi that form addition-products with dithiocarbamates and are unable to metabolize the addition products detoxify the fungicides by the storage mechanisms.

## Degradation of the Toxicant

The decomposition of a fungicide by fungi can proceed by several chemical reactions which involve metal precipitation, addition reactions, reduction, and oxidation of the toxicant.

*Metal Precipitation.* Fungi-producing materials that react with and tie up heavy metals can resist the toxic action of those metals. Oxalic acid-producing fungi convert copper to the innocuous copper oxalate (RABANUS, 1939). ASHWORTH and AMIN (1964) suggested that the capacity to produce thiols is a mechanism of metal-resistance in fungi. The mercury-sensitive fungi *Rhizoctonia solani* and *Pythium ultimum* had $^1/_{10}$ (or less) the amount of free thiols in mycelium as the mercury-resistant *Aspergillus niger*. Resistance varied with the state of reduction of sulfur nutrient, which, in turn, determined the level of protein-free thiols in the mycelium. The genetically controlled resistance of fungi to heavy metals is due, in part, to the ability of the organism to synthesize systems to maintain thiol-enriched cells (ASHIDA, 1965).

The organo and the organic parts of mercurials may be degraded by fungi. SPANIS, MUNNECKE and SOLBERG (1962) reported that several species of *Penicillium* and *Aspergillus*, conditioned to withstand normally lethal dosages of 2-chloro-4-hydroxy mercuriphenol, failed to build up resistance to methyl mercury dicyandiamide. However, with several species of *Bacillus* isolated from soil, the reverse was the case. The authors suggested that the mercurials were biodegraded with possible utilization of the organic radicals for growth.

*Secondary Addition Reactions.* The chemical reagent type fungicides are degraded by cellular compounds containing thiol and amino groups. The capacity of the fungus to produce thiols and amines in large amounts is a mechanism of resistance to fungicides.

Dichlone, on losing a chlorine atom, is capable of forming addition-products. Using dichlone-$C^{14}$, OWENS and MILLER (1957) found 75% of the sublethal dose of the fungicide taken up by conidia of *Neurospora sitophila* is bound to soluble proteins on fractionating the cells. The reaction is rapid. Little intact dichlone can be recovered from treated cells. Much of the bound dichlone is not at sites of toxicity. Dichlone reacting with cellular components not involved in toxicity is detoxified. Excessive detoxication is one factor that contributes to the large ED-50 value of dichlone.

The alkylating reactions of cellular metabolites with dyrene (BURCHFIELD and STORRS, 1956) represent mechanisms of detoxication of the fungicide. Presumably, the fungicide combines irreversibly with cellular components because less than 25% of the intact fungicide could be recovered in extractions of cells immediately following a subtoxic treatment of the fungicide. Evidently, most of the fungicide is consumed in indiscriminate reactions because an external supply of intact fungicide is required for lethal action.

The formation of an addition-product resembling a glucoside from chloroneb by *Neurospora crassa* may be a mechanism of detoxication by the fungus (HOCK, 1968). *Neurospora crassa* is highly resistant to chloroneb. Hydrolysis of one methoxy group of chloroneb by *Rhizoctonia solani* gives rise to the nontoxic 2,5-dichloro-4-methoxyphenol. *Rhizoctonia solani* is sensitive to chloroneb and hydrolysis of the fungicide represents a fraction of the uptake dose. Whether hydrolysis of a methoxy group is an intermediate step in addition-product formation in *N. crassa* is not clear. However, detoxication may not be an exclusive basis for fungous resistance to chloroneb because no detectable metabolite of the fungicide could be found in treated cultures of *Saccharomyces pastorianus*. The yeast is highly resistant to chloroneb.

*Reduction.* The reduction of captan and related R–$SCCl_3$ compounds by thiols as illustrated in Chapter 6 gives rise to the free imide, thiophosgene, and hydrochloric acid. The release of the free imide and formation of thiophosgene *in vitro* and *in vivo* are well documented [LUKENS, 1963 (2); LUKENS and SISLER, 1958 (1); OWENS and BLAAK, 1960 (2); SIEGEL and SISLER, 1968 (1). No intact captan has been recovered from cells treated with fungitoxic dosages of the fungicide (OWENS and NOVOTNY, 1959; RICHMOND and SOMERS, 1962). Decomposition proceeds *in vivo* predominantly near the cell surface. In studies using radioactive folpet, essentially all of the imide could be recovered free and intact from the ambient fluid (SIEGEL and SISLER, 1967). Pretreatment of fungous cells with sublethal amounts of thiol reagents increases the potency of captan (LUKENS and RICH, 1959; RICHMOND and SOMERS, 1962). Evidently, the reaction of captan and folpet with non-protein thiols of the cell is a means of biodetoxication of the fungicide.

An examination of the reaction products of the trichloromethylthio group of $S^{-35}$-folpet showed that 77% of the sulfur component of the fungicide escapes as a gas and 23% of the sulfur component of the fungicide accumulates in yeast cells (SIEGEL and SISLER, 1967). A part of the accumulated sulfur component may be bound to proteins [SIEGEL and SISLER, 1968 (2)]. Only a small portion of bound sulfur may actually represent fungicide reacting at the sites of action. Evidently, only a small portion of folpet or captan taken up by fungous cells reaches the sites of action. The major portion is detoxified by the fungus. Data of RICHMOND and SOMERS (1962) suggest that the detoxified portion exceeds 90%.

The reduction of the nitro group of dinitro phenol may be a means of detoxication in the plant pathogen *Fusarium oxysporium* (MADHOSINGH, 1961). *Fusarium oxysporium* can tolerate more dinitrophenol than *Coprinis mecaeus*, a common wood-rotting fungus, and *C. mecaeus* does not reduce dinitrophenol to aminonitrophenols. There was no indication of nitro elimination in *F. oxysporium* as takes place in bacteria.

*Oxidation.* Numerous organic pesticides are degraded by fungi for use as a supplemental carbon source when the fungi are starved for their normal food supply. An outstanding example of fungous adaptation to utilize a fungicide is *Hormodendrum resinae*, commonly isolated from creosote-treated wood products (CHRISTENSEN *et al.*, 1942). MARSDEN (1954) found that coal tar can serve as a nitrogen source as well as a carbon source. Sporulation was proportional to the amount of coal tar in the medium. However, the fungus had only a slight edge over other fungi to tolerate certain hydrocarbon constituents of creosote.

Presumably, fungous adaptation to various aromatic fungicides is caused by an induced system that metabolizes the aromatic nucleus. Induced resistance to one of the compounds chloronitrobenzene, chlorophenol, and Botran is common to all and the resistance is genetically controlled (GEORGOPOULUS, 1963; PRIEST and WOOD, 1961). An extracellular enzyme from *Trametes versicolor* was reported by LYR (1963) to detoxify pentachlorophenol in treated wood. The nontoxic phenols tested by RICH and HORSFALL [1954 (1)] were metabolized by polyphenol oxidase from *Stemphylium sarcinaeforme* and increased melanin pigmentation in hyphae. Presumably, in certain cases, polyphenol oxidase, in addition to promoting fungitoxic quinones from phenols, may degrade certain potentially toxic phenols.

Quinones are oxidized by fungi to several compounds, which may represent intermediate steps in decomposition (WOODCOCK, 1967). The fungus may hydroxylate an *ortho* position, cause the ring to open, and hydrolize the resulting compound to a carboxylic acid in the manner of bacterial metabolism of phenol. As an alternate pathway a double hydroxylation may take place to form gentisic acids as intermediates.

Aryl oxyalkane acids are oxidized by *Aspergillus niger* (WOODCOCK, 1967). In general, oxidation proceeds by hydroxylation on the side of the aromatic ring opposite the ether linkage. When that position is occupied by a chlorine atom, the halogen is displaced to an adjacent site during the hydroxylation. In addition, the alkane carboxylic groups through $\beta$-oxidation, are decomposed to acetic or propionic acids.

The possibility of fungi employing the $\beta$-oxidative pathway to attack the large alkyl groups of surfactant fungicides has not been investigated. Surely, without the proper hydrophobic characteristic, neither dodine nor glyodine can reach its sites of action. In apple tissue, the alkyl group of dodine is shortened to two carbons to form creatine (CURRY, 1962), presumably by $\beta$-oxidation. Dodine is converted by spores of *Fusarium solani* to a compound of less toxicity that lacks characteristics of binding to fungous cells (BARTZ and MITCHELL, 1970).

## Conclusions

The fungus can act on the fungicide to delay, promote, or prevent its toxic action by storing, activating, or detoxifying it. Mechanisms of fungicide storage may proceed by processes that reduce the activity of the toxicant in the aqueous phase. Such processes are partitioning of toxicant out of the aqueous phase to the lipoid phase and curtailing activity in the aqueous phase when the fungus is under a high water stress. The fungus can activate the toxicant by releasing it from storage or from precursors and by incorporating a synthetic compound into enzymes that fail to function. Fungous reactions of detoxication include metal precipitation, formation of addition-products and reduction-oxidation reactions.

# Appendix

## Identification and Structure of Some Fungicides

*Acridine.* Uses: no practical use.

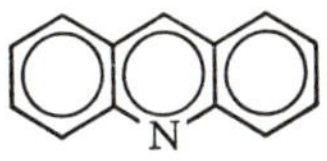

*Alkyl quaternary ammonium salts:* $R^1$-$R^4$ = H, alkyl or aryl group, x = Cl, Br. Uses: disinfectant in soaps.

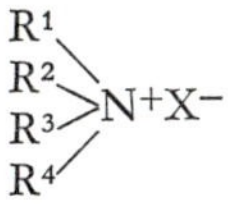

$R^1R^2R^3R^4N^+X^-$

*Ampheterícin B:* Fungizone; an antifungal polyene antibiotic including mycosamine as the nitrogen-containing moiety. Uses: to control systemic mycotic infections.

*6-Azauracil.* Uses: experimental, a systemic for leaf disease.

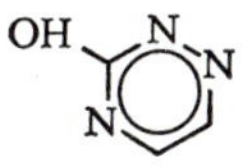

*Benlate:* 1-(butylcarbamoyl)-2-(benzimidazol) carbamic acid, methyl ester. Uses: experimental, a systemic for plant diseases.

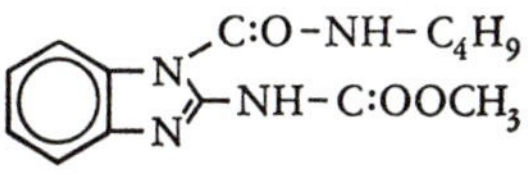

$C{:}O-NH-C_4H_9$

$NH-C{:}OOCH_3$

*Bordeaux mixture:* a mixture of copper sulfate, lime, and water (e.g. 4 lbs — 10 lbs — 100 gal). Uses: foliar protectant.

*Botran:* 2,6-dichloro-4-nitroaniline (dichloran); Uses: a post-harvest treatment to retard spoilage entransit.

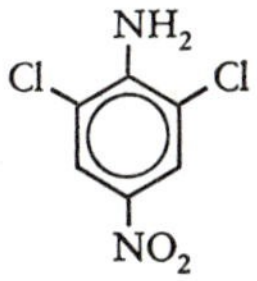

$NH_2$ Cl Cl $NO_2$

*Captan:* *N*-trichloromethylthio-4-cyclohexane-1,2-dicarboximide. Uses: foliar protectant, disinfectant in soaps.

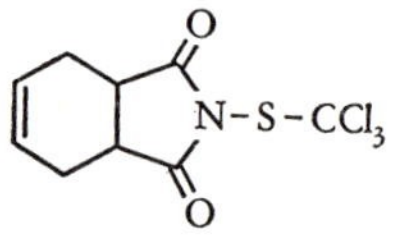

*Chloramphenicol:* D-(-)-threo-2-dichloroacetamido-1-*p*-nitrophenyl-1,3-propanediol. Uses: medical antibacterial agent.

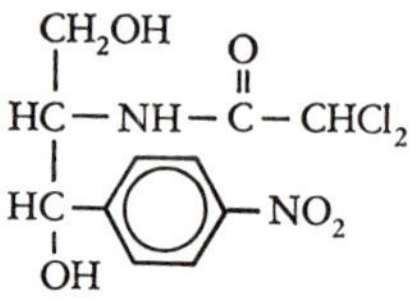

*2-Chlorohydroxymercuriphenol.* Uses: seed treatment and turf spray.

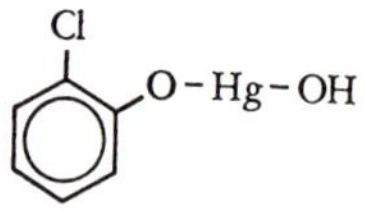

*Chlorophenol:* phenol of one to five chlorine substituents. Uses: wood preservative.

*Chloropicrin:* trichloronitromethane. Uses: fumigant for soil, plant beds, and greenhouse.

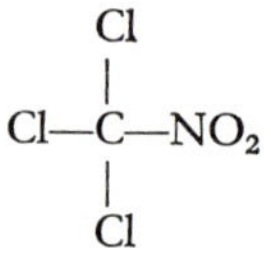

*Creosote:* a complex from coal tar, which exact structure is unknown. Uses: wood preservative.

*Cuprous oxide.* Uses: foliar protectant.

$$Cu_2O$$

*Cycloheximide:* β-[2(3,5-dimethyl-2-oxocyclohexyl)-2-hydroxyethyl]-glutarimide. Uses: foliar spray.

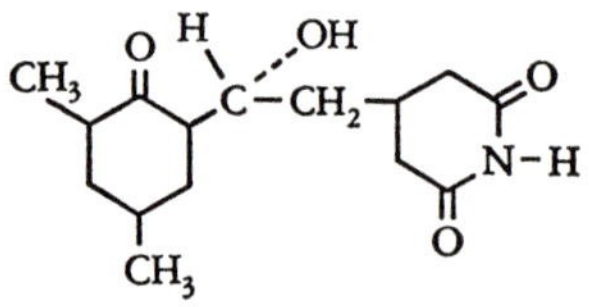

*Cycloheximide acetate.* Uses: experimental.

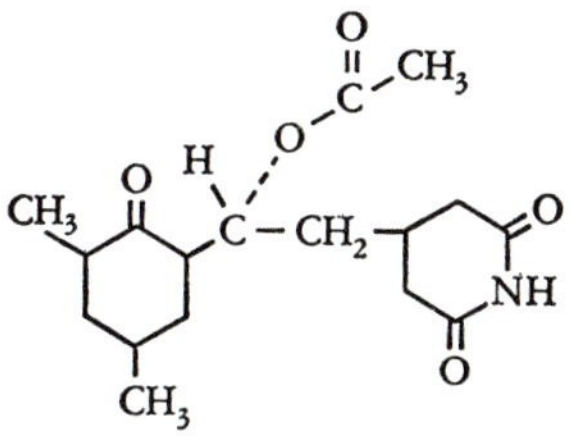

*Cycloheximide oxime.* Uses: experimental.

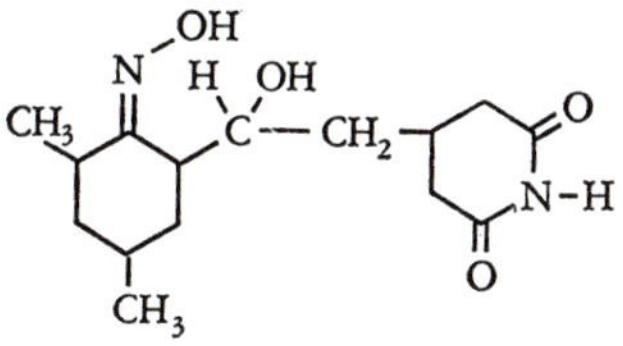

*D-D; A mixture of dichloropropanes and dichloropropenes.* Uses: soil fumigant predominantly used to control plant parasitic nematodes.

$ClCH_2$-$CH_2$-$CH_2Cl$ $ClCH_2$-CH = CHCl

*Dichlone:* 2,3-dichloro-1,4-naphthoquinone. Uses: foliar protectant and seed treatment.

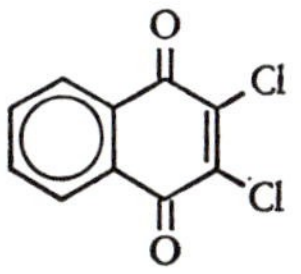

*Difolatan:* *N*-1,1,2,2-tetrachloroethylthio-4-cyclohexene-1,2-dicarboximide. Uses: foliar protectant.

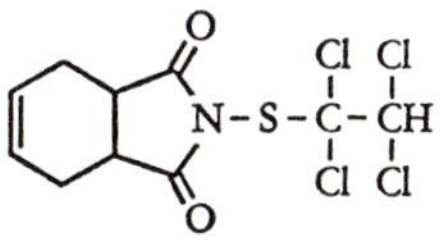

*Dihydrocarboxanilidomethyl oxathiin:* Vitavax. Uses: a systemic fungicide for seed treatment of grains against loose smuts.

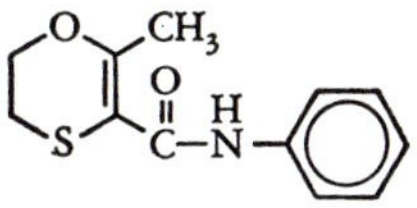

*Dihydrocarboxanilidomethyl oxathiin dioxide:* Plantvax. Uses: an experimental systemic fungicide for treatment of soil or plant against rust diseases.

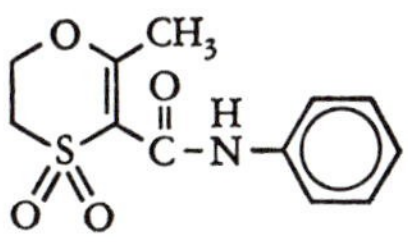

*Dodine:* dodecylguanidine acetate. Uses: foliar protectant with slight eradicating properties.

$$C_{12}H_{25}\text{-}N(H)\text{-}C(=NH)\text{-}NH_3^+ \cdot CH_3\text{-}C(=O)\text{-}O^-$$

*Dyrene:* 2,4-dichloro-6-(2-chloroanilino)-1,3,5-*s*-triazine. Uses: foliar protectant.

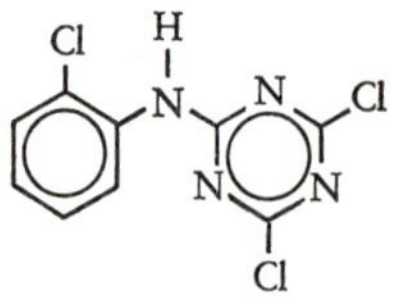

*Ethionine.* Uses: experimental systemic for plant disease.

$$C_2H_5\text{-}S\text{-}CH_2\text{-}CH(NH_2)\text{-}C(=O)\text{-}OH$$

*Ethylene dibromide.* Uses: soil fumigant.

$$Br\text{-}CH_2\text{-}CH_2\text{-}Br$$

*Ethylene-bis-dithiocarbamate:* see nabam.

*Filipin:* an antifungal agent produced by *Streptomyces filipinensis.* Its structure is considered to be a conjugated polyene.

*Folpet:* *N*-trichloromethylthiophthalimide. Uses: foliar protectant.

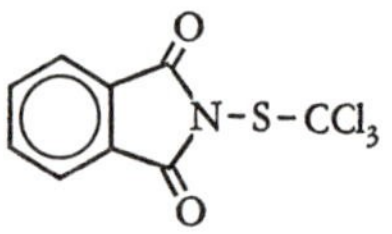

*Glyodin:* 2-heptadecylimidazoline acetate. Uses: foliar protectant.

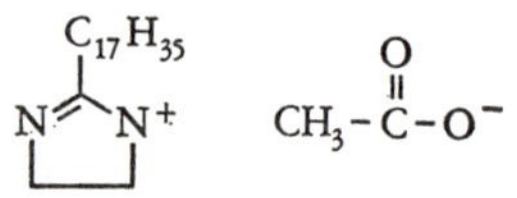

*8-Hydroxyquinoline.* Uses: metal salts are used as antimicrobial agents in material and as seed protectant.

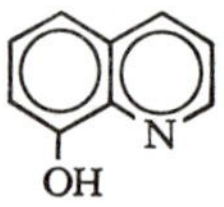

*Karathane:* 2,4-dinitro-6-(1′-methyl-heptyl)phenyl crotonate. Uses: foliar spray for powdery mildew diseases.

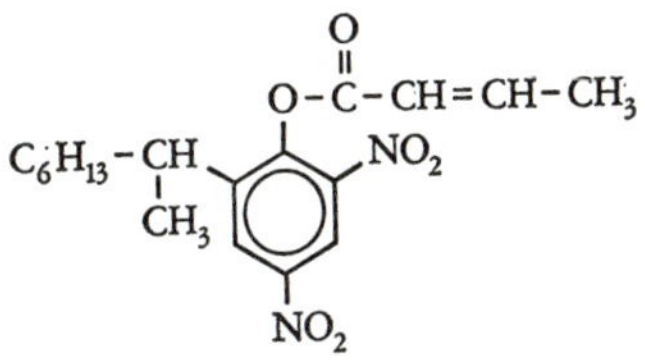

$$C_6H_{13}-CH(CH_3)-C_6H_2(NO_2)_2-O-C(=O)-CH=CH-CH_3$$

*Maneb:* manganous ethylene-bis-dithiocarbamate. Uses: foliar protectant.

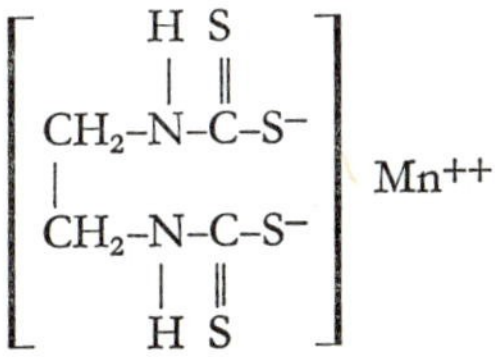

$$\left[\begin{array}{l} CH_2-NH-C(=S)-S^- \\ | \\ CH_2-NH-C(=S)-S^- \end{array}\right] Mn^{++}$$

*Mercuric chloride.* Uses: seed treatment, disinfectant, foliar eradicant for turf.

$$HgCl_2$$

*Methyl bromide.* Uses: soil fumigant.

$$CH_3Br$$

*Methyl isothiocyanate.* Uses: soil fumigant.

$$CH_3 = N = C = S$$

*Methyl mercury dicyandiamide.* Uses: Seed treatment, soil drench, foliar eradicant for turf.

$$CH_3-Hg-NH-C(=NH)-NH-C \equiv N$$

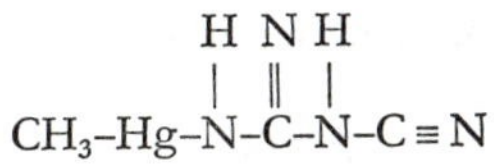

*Mitomycin C:* produced by *streptomyces caespitosus.* Structure unknown. Uses: medical antitumor agent.

*Mylone:* 3,5-dimethyl-1,3,5,2-H-tetrahydrothiazine-2-thione. Uses: soil fumigant.

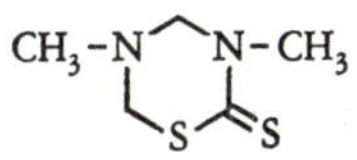

$$CH_3-N \quad N-CH_3 \;(\text{ring with } S \text{ and } C=S)$$

*Nabam:* disodium ethylene-bis-dithiocarbamate. Uses: soil drench and mixed with $Zn^{++}$ or $Mn^{++}$ in spray tank to form zineb and maneb, respectively.

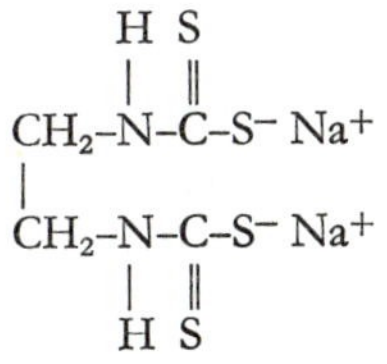

$$\begin{array}{l} CH_2-NH-C(=S)-S^-\,Na^+ \\ | \\ CH_2-NH-C(=S)-S^-\,Na^+ \end{array}$$

*Pentachloronitrobenzene.* Uses: soil fungicide.

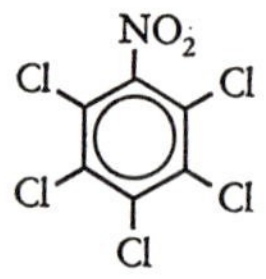

*Phenyl mercury acetate.* Uses: turf spray and industrial fungicide.

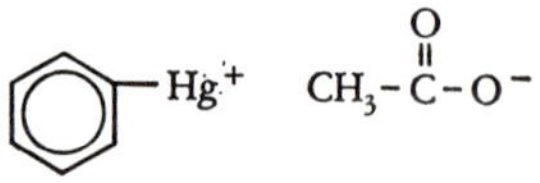

*o-Phenyl phenol:* Uses. Industrial fungicide.

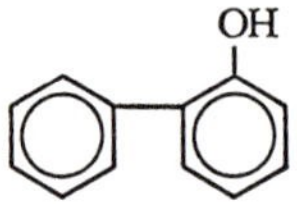

*Puromycin*: 6-dimethylamino-9-[3-deoxy-3-(*p*-methoxy-L-phenyl-alanyl amino)-*B*-D-ribofuranosyl]-*B*-purine. Uses: experimental antibacterial antibiotic.

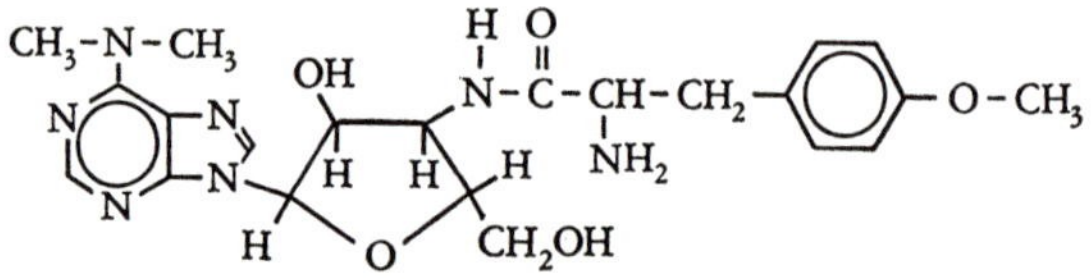

*2-Pyridinethione-N-oxide.* Uses: disinfectant in soaps.

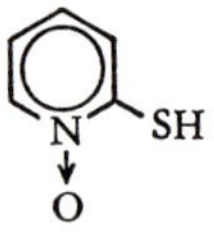

*Sulfur. S and polysulfides.* Uses: foliar protectant.

*Sulfur dioxide.* Uses: fumigant for storage bins.

$$O=S=O$$

*Tetrachloroisophthalonitrile* (*Daconil 2787*). Uses: foliar protectant.

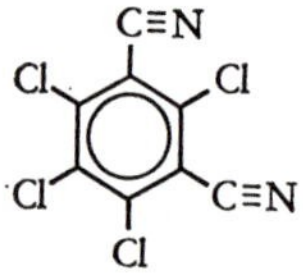

*Thiram:* tetramethylthiuram disulfide. Uses: seed treatment and foliar protectant.

$$(CH_3)_2N-C(=S)-S-S-C(=S)-N(CH_3)_2$$

*Triphenyltin acetate.* Uses: experimental foliar protectant.

$$(C_6H_5)_3Sn^+ \quad CH_3C(=O)O^-$$

*Vapam:* sodium methyl dithiocarbamate. Uses: soil sterilant.

$$CH_3-NH-C(=S)-S^- \; Na^+$$

*Zineb:* zinc ethylene-bis-dithiocarbamate. Uses: foliar protectant.

$$\left[\begin{array}{l} CH_2-NH-C(=S)-S^- \\ | \\ CH_2-NH-C(=S)-S^- \end{array}\right] Zn^{++}$$

*Ziram:* zinc dimethyldithiocarbamate. Uses: foliar protectant.

$$\left[(CH_3)_2N-C(=S)-S^-\right]_2 \; Zn^{++}$$

# References


ABRAMS, E.: Microbiological deterioration of organic material. Natl. Bur. Standards (U.S.) Misc. Publ. **188**, 1—41 (1948).

ADAM, N. K.: Water repellency. Endeavour **17**, 37—41 (1958).

— The chemical structure of solid surfaces as deduced from contact angles. In: Advances in Chem. Ser. No. 43, p. 52—56. (GOULD, R. F., Ed.). Contact angle, wettability, and adhesion. American Chemical Society 1964.

ALBERT, A.: Kationic chemotherapy with special reference to the acridines. Med. J. Australia **1**, 245—249 (1944).

— Selective toxicity, 2. ed., p. 83—187. London: Methuen and Co. LTD 1960.

— GIBSON, M. I., RUBBO, S. D.: The influence of chemical constitution on antibacterial activity. VI. The bacterial action of 8-hydroxyquinoline (oxine). Brit. J. Exptl. Pathol. **34**, 119—138 (1953).

ANDERSON, B. M., REYNOLDS, M. L.: Inhibition of yeast alcohol dehydrogenase by alkylammonium chlorides. Biochim. et Biophys. Acta **96**, 45—51 (1965).

ANDREEVA, E. I., MEL'NIKOV, N. N., SAPOZHKOV, YU. N., SHVETSOVA-SHILOVSKAYA, K. D.: The fungicidal activity of some organic arsenic compounds. Khim. Sel'sk. Khoz. **3**, 28—30 (1965).

ANON: Definition of fungicide terms. Phytopathology **33**, 624—626 (1943).

— Sourcebook of laboratory exercises in plant pathology, p. 509. American Phytopathological Society (mimeograph copy), 1962.

— Fungicidal dialkylthiono thiolophosphoric (or phosphonic) acid esters. Farbenfabriken Bayer A. G. Neth. Appl. **6**, 237, 406 (1964).

— Fungicidal isocyanide dihalides. Farbenfabriken Bayer A. G. Brit. **981**, 107 (1965).

ARK, P. A., WILSON, E. M.: Availability of streptomycin in dust formulations. Plant Disease Reptr **40**, 332—334 (1956).

ARONSON, J. M.: The cell wall. In: The fungi, an advanced treatise. Vol. I., 1965. The fungal cell, Chapter 3, 47—76. (AINSWORTH, G. C., SUSSMAN, A. S., Eds). New York: Academic Press Inc.

ASHIDA, J.: Adaptation of fungi to metal toxicants. Ann. Rev. Phytopath. **3**, 153—174 (1965).

ASHWORTH, L. J., AMIN, J. V.: A mechanism for mercury tolerance in fungi. Phytopathology **54**, 1459—1463 (1964).

BAECHLER, R. H.: Toxicity of normal aliphatic alcohols, acids, and sodium salts. Am. Wood-Preservers' Assoc. **35**, 1—8 (1939).

— Application and use of fungicides in wood preservation. In: Fungicides, an advanced treatise, Vol. 1, Chapter 11, 425—461. (TORGESON, D. C., Ed.). New York City: Academic Press, Inc. 1967.

BAKER, B. R.: Interactions of enzymes and inhibitors. J. Chem. Educ. **44**, 610—619 (1967).

BÁNHIDI, Z. G.: Activation of the disulfide forms of thiamine and its phosphate as growth factors for Lactobacillus fermenti. J. Bacteriol. **79**, 181—190 (1960).

BARRON, E. S. G., ARDO, M. I., HEARON, M.: Regulatory mechanisms of cellular respiration. III. Enzyme distribution in the cell. Its influence on the metabolism of pyruvic acid by baker's yeast. J. Gen. Physiol. **34**, 211—224 (1950).

BARTNICKI-GARCIA, S.: Cell wall chemistry, morphogenesis, and taxonomy of fungi. Ann. Rev. Microbiol. **22**, 87—108 (1968).

BARTZ, J. A., MITCHELL, J. E.: Comparative interaction of n-dodecylquanidine acetate with four plant pathogenic fungi. Phytopathology **60**, 345—349 (1970).

BENNETT, R. F.: Biologically active chemicals: development of organotins. Rubber & Plastics Age. **46**, 260—261 (1965).

BENT, K. J., MOORE, R. H.: The mode of action of griseofulvin. Biochemical studies of antimicrobial drugs. 16th Symposium of Soc. Gen. Microbiol., p. 82—110, London, 1966. Cambridge: University Press 1966.

BERARD, W. N., LEONARD, E. K., REEVES, W. A.: Cotton made resistant to microbiological deterioration using formic acid colloid of methylolmelamine. Develop. Ind. Microbiol. **2**, 79—91 (1961).

BLANK, H., TAPLIN, D., ROTH, F. J.: Electron microscopic observations of the effects of griseofulvin on dermatophytes. Arch. Dermatol. and Syphilol. **81**, 667 (1960).

BLISS, C. I.: The calculation of the dosage-mortality curve. Ann. Appl. Biol. **22**, 134—167 (1935).

— Some principles of bioassay. Am. Scientist **45**, 449—466 (1957).

BLOCK, S. S.: Fungitoxicity of the 8-quinolinols. J. Agr. Food Chem. **3**, 229—234 (1955).

— Reversal of fungitoxicity of copper-8-quinolinolate. J. Agr. Food Chem. **4**, 1042—1046 (1956).

— Fungicides as industrial preservatives. In: Fungicides, an advanced treatise, Vol. I, Chapter 10, 379—423. (TORGESON, D. C., Ed.). New York City: Academic Press 1967.

BODNAR, J., TERENYI, A.: Biochemistry of the smut diseases of cereals. Note 4. The mechanism of the action of mercury salts on the spores of wheat bunt. (Tilletia tritica [Bjerk] Winter). Z. physiol. Chem., Hoppe-Seyler's **207**, 78—92 (1932).

BOMAR, M., JEDLINSKI, Z., HIPPE, R.: Antifungal activity of some 3-phenoxy-1,2-epoxypropane derivatives. Folia Microbiol (Prague) **11**, 155—158 (1966).

BRIAN, P. W.: Griseofulvin. Trans. Brit. Mycol. Soc. **43**, 1—13 (1960).

BROCKMAN, R. W., ANDERSON, E. P.: Pyrimidine analogues. In: Metabolic inhibitors, Vol. 1, 239—285. (HOCHSTOR, R. M., QUASTEL, J. H., Eds). New York: Academic Press 1963.

BROOK, M. Differences in the biological activity of 2,3,5,6-tetrachloronitrobenzene and its isomers. Nature **170**, 1022 (1952).

BROOK, P. J.: A comparison of glasshouse and laboratory methods for testing fungicides against Botrytis cinera. New Zealand J. Sci. Technol. A. **38**, 506—511 (1957).

BROWN, I. F., SISLER, H. D.: Mechanisms of fungitoxic action of n-dodecylguanidine acetate. Phytopathology **50**, 830—839 (1960).

BRUNSKILL, R. T.: Physical factors affecting the retention of spray droplets on leaf surfaces. Brit. 3rd Weed Control Conf., Blackpool 1956. Proc. pg. 593—603 (1956).

BURACZEWSKI, K., CZERWINSKA, E., ECKSTEIN, Z., GROCHOWSKI, E., KOWALIK, R., PLENKIEWICZ, J.: The properties and fungicidal activity of some aryl derivatives of hydroxyamic acid. Przemysl Chem. **43**, 626—629 (1964).

BURCHFIELD, H. P.: Comparative stabilities of dyrene, 1-fluoro-2,4-dinitrobenzene, dichlone and captan in a silt loam soil. Contribs Boyce Thompson Inst. **20**, 205—215 (1959).

— Chemical and physical interactions. In: Fungicides, an advanced treatise, Vol. 1, Chapter 12, p. 463—508. (TORGESON, D. C., Ed.). New York City: Academic Press 1967.

— GOENAGA, A.: Some factors governing the deposition and tenacity of copper fungicide sprays. Contribs Boyce Thompson Inst. **19**, 141—156 (1957).

— McNEW, G. L.: Mechanism of particle size effects of fungicides on plant protection. Contribs Boyce Thompson Inst. **16**, 131—161 (1950).

— SCHECHTMAN, JOAN: Absorptiometric analysis of *N*-(trichloromethylthio)-4-cyclohexene-1,2-dicarboximide (captan). Contribs Boyce Thompson Inst. **19**, 411—461 (1958).

— — MAGDOFF, BEATRICE: Effects of temperature and composition on crystallite growth in Bordeaux mixture. Contribs Boyce Thompson Inst. **19**, 117—132 (1957).

— STORRS, ELEANOR: Chemical structures and dissociation constants of amino acids, peptides, and protein in relation to their reaction rates with 2,4-dichloro-6-(*o*-chloroanilino)-2-triazine. Contribs Boyce Thompson Inst. **18**, 395—418 (1956).

— — Effect of chlorine substitution and isomerism on the interactions of s-triazine derivatives with conidia of Neurospora sitophila. Contribs Boyce Thompson Inst. **18**, 429—452 (1957).

— — Relative reactivities of 1-fluoro-2,4-dinitrobenzene and 2,4-dichloro-6(*o*-chloroanilino)-*s*-triazine with metabolites containing various functional groups. Contribs Boyce Thompson Inst. **19**, 169—176 (1958).

Byrde, R. J. W., Clifford, D. R., Woodcock, D.: Fungicidal activity and chemical constitution. XIII. Active components of commercial formulations containing dinocap. Ann. Appl. Biol. **57**, 223—230 (1966).
— Martin, J. T., Nicholas, D. J. D.: Effects of fungicides on fungus enzymes. Nature **178**, 638—639 (1956).
— Woodcock, D.: Fungicidal activity and chemical constitution II. Compounds related to 2,3-dichloro-1,4-naphthaquinone. Ann. Appl. Biol. **40**, 675—687 (1953).
— — Fungicidal activity and chemical constitution. VII, a study of the structure specificity of seven fungicides. Ann. Appl. Biol. **47**, 332—338 (1959).
Campaigne, E., Tsurugi, J., Meyer, W. W.: The ultraviolet absorption spectra of some unsymmetrical disulfides. J. Org. Chem. **26**, 2486—2491 (1961).
Carson, R.: Silent spring, pp. 5—37. Boston, Mass.: Houghton 1962.
Cassie, A. B. D.: Discussions. Faraday Soc. **3**, 14 (1948).
Cheymol, J., Charbier, P., Seyden-Penne, J., Moreau, Cl., Moreau, M.: New derivatives of pentaerythritol with fungistatic activity. Ann. pharm. franç. **22**, 595—602 (1964).
Christensen, C. M., Kaufert, F. H., Schmitz, H., Allison, J. L.: Hormodendrum resinae (Lindau), an inhabitant of wood impregnated with creosote and coal tar. Am. J. Botany **29**, 552—558 (1942).
Clark, N. G., Hans, A. F.: Antifungal activity of substituted nitroaniline and related compounds. J. Sci. Food Agr. **12**, 751—757 (1961).
Cochrane, V. W.: Physiology of fungi, p. 202—232. New York-London: J. Wiley and Sons, Inc. 1958.
Collander, R.: The permeability of Nitella cells to non-electrolytes. Physiol. Plantarum **7**, 420—445 (1954).
Colombo, B., Felicetti, L., Baglioni, C.: Inhibition of protein synthesis by cycloheximide in rabbit reticulocytes. Biochem. Biophys. Res. Commun. **18**, 389—395 (1965).
Commager, H., Judis, J.: Mechanism of action of phenolic disinfectants. VI. Effects of glucose and succinate metabolism of Escherichia coli. J. Pharm. Sci. **54**, 1436—1439 (1965).
Cooke, J. R.: A probabilistic analysis of fungus control by foliar application of chemicals. Ph. D. Thesis, North Carolina State University at Raleigh, 100 p., 1966.
Coursen, B. W., Sisler, H. D.: Effect of the antibiotic, cycloheximide, on the metabolism and growth of Saccharomyces pastorianus. Am. J. Botany **47**, 541—549 (1960).
Courshee, R. J.: Application and use of foliar fungicides. In: Fungicides, an advanced treatise, Vol. 1, Chapter 8, p. 239—286. (Torgeson, D. C., Ed.). New York City: Academic Press 1967.
— Ireson, M.: The range of projection of small drops. J. Agr. Eng. Research **6**, 59—63 (1961).
Crowdy, S. H.: Uptake and translocation of organic chemicals by higher plants. Plant Pathology, Problems and Progress, Chapter 22, 231—238, 1908—1958. Madison: Univ. Wisconsin Press 1959.
Curry, A. N.: Translocation and metabolism of dodecylguanidine acetate (dodine) fungicide in apple trees using $C^{14}$ radio tagged dodine. J. Agr. Food Chem. **10**, 13—18 (1962).
Czerkowicz, A. B., Stuart, L. S.: Methods of testing fungicides. In: Disinfection, sterilization and preservation, pp. 207—217. (Lawrence, C. A., Block, S. S., Eds). Philadelphia: Lea and Febiger 1968.
Daines, R. H., Brennan, E., Leone, I. A.: Effect of plant bed temperature and seed potato dip treatments on incidence of sweet potato sprout decay caused by Diaporthe batatatis. Phytopathology **50**, 186—187 (1960)
— Lukens, R. J., Brennan, E., Leone, I. A.: Phytotoxicity of captan as influenced by formulation, environment, and plant factors. Phytopathology **47**, 567—572 (1957).
Davson, H., Danielli, J. F.: The permeability of natural membranes, pp. 60—75. Cambridge: Cambridge University Press 1943.
Dekker, J., Oort, A. J. P.: Mode of action of 6-azauracil against powdery mildew. Phytopathology **54**, 815—818 (1964).
— Tulleners, I.: The antibiotic griseofulvin, some aspects of its mode of action. Mededeel. Landbouwhogeschool Opzoekingssta. Staat Gent. **28**, 574—579 (1963).

DELP, C. J., KLÖPPING, H. L.: Performance attributes of a new fungicide and mite ovicide candidate. Plant Disease Reptr **52**, 95—99 (1968).

DENISKINA, E., DRUI, G., KHOKHRYAKOVA, V. S., KOGAN, L. M.: Action of hexachlorobutadiene and polychlorobutanes on pure cultures of some soil microorganisms. Khim v Sel'sk. Khoz. **4** (2), 28—30 (1966).

DIMOND, A. E.: Surface factors affecting the penetration of compounds into plants. In: Modern methods of plant analysis, Vol. V, 368—382. (LINSKENS, M. F., TRACEY, M. V., Eds). Berlin-Göttingen-Heidelberg: Springer 1962.

— DUGGER, B. M.: Some lethal effects of ultraviolet radiation on fungous spores. Proc. Natl. Acad. Sci. US **27**, 459—468 (1941).

— HEUBERGER, J. W., HORSFALL, J. G.: A water soluble protectant fungicide with tenacity. Phytopathology **33**, 1097 (1943).

— HORSFALL, J. G., HEUBERGER, J. W., STODDARD, E. M.: Role of the dosage-response curve in the evaluation of fungicides. Conn. Agr. Expt. Sta., New Haven, Bull. **451**, 635—667 (1941).

— STODDARD, E. M.: Toxicity to greenhouse roses from paints containing mercury fungicides. Conn. Agr. Expt. Sta., New Haven, Bull. **595**, 1—19 (1955).

DOMSCH, K. H.: Der Einfluß von fungiziden Wirkstoffen auf die Bodenatmung. Phytopathol. Z. **49**, 291—302 (1964).

DOUGLAS, H. W., COLLINS, C. E., PARKINSON, D.: Electric charge and other surface properties of some fungal spores. Biochim. et Biophys. Acta. **33**, 535—538 (1959).

DUGGER, W. M. JR., HUMPHREYS, T. E., CALHOUN, B.: Influence of N-(trichloromethylthio)-4-cyclohexene-1,2-dicarboximide (captan) on higher plants. II. Effect of specific enzyme systems. Am. J. Botany **46**, 151—156 (1959).

DUTHIE, E. S.: The production of penicillinase by organisms of the subtilis group. Brit. J. Exptl Pathol. **25**, 96—100 (1944).

DUYFJES, W.: The formulation of pesticides. Philips Tech. Rev. **19**, 165—176 (1958).

ECKERT, J. W.: Fungistatic and phytotoxic properties of some derivatives of nitrobenzene. Phytopathology **52**, 642—649 (1962).

— Application and use of post harvest fungicides. In: Fungicides, an advanced treatise, Vol. I, Chapter 9, 287—378. (TORGESON, D. C., Ed.). New York City: Academic Press 1967.

ECKERT, J. M., KOLBENZEN, M. J., BRETSCHNEIDER, B. F., NICHOLAS, H. K.: Controlling penicillium decay of oranges with certain aliphatic amines. Phytopathology **51**, 64 (1961).

EDGINGTON, L. V.: Effect of chain of the alkyl quaternary ammonium compounds upon their use as systemic fungicides. Phytopathology **56**, 23—25 (1966).

— BARRON, G. L.: Fungitoxic spectrum of oxathiin compounds. Phytopathology **57**, 1256—1257 (1967).

— DIMOND, A. E.: The effect of adsorption of organic cations to plant tissue on their use as systemic fungicides. Phytopathology **54**, 1193—1197 (1964).

— WALTON, G. S., MILLER, P. M.: Fungicide selective for Basidiomycetes. Science **153**, 307—308 (1966).

EL-ZAYAT, M. M., LUKENS, R. J., DIMOND, A. E., HORSFALL, J. G.: Systemic action of nitrophenols against powdery mildew. Phytopathology **58**, 434—437 (1968).

EMMONS, C. W.: Chemotherapeutic and toxic activity of antifungal agent X-5079C in experimental mycoses. Am. Rev. Resp. Dis. **84**, 507—511 (1961).

ESPOSITO, R. G., FLETCHER, ALISON M.: The relation of pteridine biosynthesis to the action of copper 8-quinolinolate on fungal spores. Arch. Biochem. Biophys. **92**, 369—376 (1961).

ESSER, K., KUENEN, R.: Genetics of fungi, p. 340—438. Berlin-Heidelberg-New York: Springer 1967.

EVANS, A. C., MARTIN, H.: The incorporation of direct with protective insecticides and fungicides. I. The laboratory evaluation of water-soluble wetting agents as constituents of combined washes. J. Pomol. Hort. Sci. **13**, 261—292 (1935).

EVANS, E., COX, J. R., TAYLOR, J. W. H., RUNHAM, R. L.: Some observations on size and biological activity of spray deposits produced by various formulations of copper oxychloride. Ann. Appl. Biol. **58**, 131—144 (1966).

— Taylor, J. W. H., Runham, R., McLain, M.: A comparison of certain physical characteristics of Bordeaux mixture and a copper oxychloride wettable powder, and their possible significance in relation to biological performance. Trans. Brit. Mycol. Soc. **45**, 81—92 (1962).

Evans, G., White, N. H.: Effect of the antibiotic radicicolin and griseofulvin on the fine structure of fungi. J. Exptl Botany **18**, 465—470 (1967).

Eyring, H.: Untangling biological reactions. Science **154**, 1609—1613 (1966).

Fergeson, J., Pirie, H.: The toxicity of vapours to the grain weevil. Ann. Appl. Biol. **35**, 532—550 (1948).

Finney, D. J.: Statistical method in biological assay, 2nd, 668 p. London: Charles Griffin and Comp. Ltd 1964.

Flor, H. H.: The fungicidal activity of furfural. Iowa State College J. Sci. **1**, 199—223 (1927).

Fogg, G. E.: The penetration of 3,5-dinitro-*o*-cresol into leaves. Ann. Appl. Biol. **35**, 315—330 (1948).

Franklin, T. J. Mode of action of the tetracyclines. Biochemical Studies of Antimicrobial Drugs. 16th Symposium of Soc. Gen. Microbiol., pp. 92—212, London 1966. Cambridge: Univ. Press 1966.

Frear, D. E. H.: Chemistry of insecticides and fungicides. pp. 153—180. New York: D. van Nostrand Co., Inc. 1942.

Fujita, T., Iwasa, J., Hansch, C.: A new substituent constant $\pi$, derived from partition coefficients. J. Am. Chem. Soc. **86**, 5175—5180 (1964).

Fulton, R. H.: Low-volume spraying. Ann. Rev. Phytopathology **3**, 175—196 (1965).

Furmidge, G. G. L.: Physio-chemical studies on agricultural sprays. IV. The retention of spray liquids on leaf surfaces. J. Sci. Food Agr. **13**, 127—140 (1962).

Gaddum, J. H.: Reports on biological standards. III. Methods of biological assay depending on a quantal response. Privy Council. Medical Res. Coin. Spec. Rept. Ser. 183. 46 p. (1933).

Gale, G. R.: Cytology of Candida albicans as influenced by drugs acting on the cytoplasmic membrane. J. Bacteriol. **86**, 151—157 (1963).

— McLain, H. H.: Effect of thiobenzoate on cytology of Candida albicans. J. Gen. Microbiol. **36**, 297—301 (1964).

Gentles, J. C.: Experimental ringworm in guinea pigs: oral treatment with griseofulvin. Nature **182**, 476—477 (1958).

Georgopoulos, S. G.: Tolerance to chlorinated nitrobenzenes in Hypomyces solani f. cucurbitae and its mode of inheritance. Phytopathology **53**, 1086—1093 (1963).

— Zafiratos, C., Georgiadis, E.: Membrane functions and tolerance to aromatic hydrocarbon fungitoxicants in conidia of Fusarium solani. Physiol. Plant. **20**, 373—381 (1967).

Geronimus, L. H., Cohen, S.: Induction of staphylococcal penicillinase. J. Bacteriol. **73**, 28—34 (1957).

Gershon, H., Parmegiani, R.: Secondary mechanisms of antifungal action of substituted 8-quinolinols. I. 5- and 5,7-substituted 8-methoxyquinolines. Contribs Boyce Thompson Inst. **24**, 33—36 (1968).

Goksør, J.: The effect of some dithiocarbamyl compounds on the metabolism of fungi. Physiol. Plantarum **8**, 719—835 (1955).

Goldsworthy, M. C., Green, E. L.: Availability of the copper of Bordeaux mixture residues and its absorption by the conidia of Sclerotinia fructicola. J. Agr. Research **52**, 517—533 (1936).

Gottlieb, D., Carter, M. E., Wu, L., Sloneker, J. H.: Inhibition of fungi by filipin and its antagonism by sterols. Phytopathology **50**, 594—603 (1960).

Götz, H.: Advances in the treatment of mycoses. In: Recent advances of human and animal mycology, pp. 357—368. Bratislava: The Slovak Acad. Sci. 1967.

Granin, E. F., Bliznyuk, N. K., Levskaya, G. S., Zhil'tsova, G. I., Matyukhina, E. N., Vrublevskaya, L. S.: Fungus toxicity of some compounds of arsenic. Khim. v. Sel'sk. Khoz. **3** (2), 26—30 (1965).

Gray, R. A.: Increasing the absorption of streptomycin by leaves and flowers with glycerol. Phytopathology **46**, 105—111 (1956).

Gregory, P. H.: Deposition process and natural deposition. In: The microbiology of the atmosphere, 58—89. New York City: Interscience 1961.

GROLLMAN, A. P.: Structural basis for inhibition of protein synthesis by emetine and cycloheximide based on an analogy between ipecac alkaloids and glutarimide antibiotics. Proc. Natl Acad. Sci. (US) **56**, 1867—1874 (1966).

GROVER, R. K.: Effect of some fungicides on pectolytic enzyme activity of Sclerotinia sclerotiorum and Botrytis allii. Phytopathology **54**, 130—133 (1964).

— MOORE, J. D.: Toximetric studies of fungicides against the brown rot organisms, Sclerotinia fructicola and S. laxa. Phytopathology **52**, 876—880 (1962).

GRUENHAGEN, R. H., WOLF, P. A., DUNN, E. E.: Phenolic fungicides in agriculture and industry. Contribs Boyce Thompson Inst. **16**, 349—356 (1951).

HALL, A. N., LEESON, J. A., RYDON, H. N., TWEDDLE, J. C.: The degradation of some Bz-substituted tryptophans by Escherichia coli tryptophanase. Biochem. J. **74**, 209—216 (1960).

HAMILTON, J. M.: Evaluation of fungicides in the greenhouse. In: Plant pathology, problems and progress, 1908—1958, p. 253—257. (HOLTON, C. S., Ed.). Madison: The University of Wisconsin Press 1958.

— Redistribution of fungicides on apple leaves. First Intern. Cong. Plant Pathology, London. Abstract. **78** (1968).

— MACK, G. L., PALMITER, D. H.: Redistribution of fungicide on apple foliage. Phytopathology **33**, 5 (Abstract) (1943).

HAMILTON-MILLER, J. M. T.: Effect of EDTA upon bacterial permeability to benzylpenicillin. Biochem. Biophys. Res. Commun. **20**, 688—691 (1965).

HAMS, A. F., COLLYER, J., HOUSLEY, J. R.: Antifungal compositions. Brit. Patent 999,802 (1965). [Chem. Abstracts **63**, 12260 (1965)].

HANSCH, C., FUJITA, T.: $\varrho$-$\sigma$-$\pi$ analysis. A method for the correlation of biological activity and chemical structure. J. Am. Chem. Soc. **86**, 1616—1626 (1964).

— STEWARD, A. RUTH, IWASA, JUNKICHI, DEUTSCH, EDNA W.: The use of a hydrophobic bonding constant for structure-activity correlations. Mol. Pharmacology **1**, 205—213 (1965).

HARTILL, W. F. T.: The distribution of fungal spores and spray deposits on field tobacco. First Intern. Cong. Plant Pathology, London. Abstracts 80 (1968).

HASHIOKA, Y., TAKAMURA, Y.: Phytopharmacological studies on the rice diseases. IV. Protoplasmic alteration of conidia of Cochliobolus miyabeanus in the organomercurial solutions. Research Bul. Gifu Univ. **7**, 49—54 (1956).

HAY, R. W.: Some reactions of coordinated ligands containing oxygen and nitrogen donors. Chem. Educ. **42**, 413—417 (1965).

HEMWALL, J. B.: Theoretical considerations of several factors influencing the effectivity of soil fumigants under field conditions. Soil Sci. **90**, 157—168 (1960).

HICOCK, H. W., OLSON, A. R., CALLWARD, F. M.: Preserving wood for farm use. Univ. Connecticut Extension Bull (Storrs) **415**, 1—20 (1949).

HIGUCHI, T.: Physical chemical analysis of percutaneous absorption process from creams and ointments. J. Soc. Cosmetic Chem. **11**, 85—95 (1960).

HILDICK-SMITH, G., BLANK, H., SARKANY, I.: Fungus diseases and their treatment, pp. 367 to 486. Boston: Little, Brown, and Co. 1964.

HILL, E. P.: Uptake and translocation. In: The fungi, an advance treatise. Vol. I. Chapter **16**, 457—463. (AINSWORTH, G. C., SUSSMAN, A. S., Eds). New York: Academic Press 1965.

HILLS, F. J., LEACH, L. D.: Photochemical decomposition and biological activity of *p*-dimethylaminobenzenediazo sodium sulfonate (Dexon). Phytopathology **52**, 51—56 (1962).

HINE, J.: Physical organic chemistry, pp. 66—80, 123—185. New York: McGraw-Hill Book Comp. 1962.

HOCHSTEIN, P. E., COX, C. E.: Studies on the fungicidal action of *N*-(trichloromethylthio)-4-cyclohexene-1,2-dicarboximide (captan). Am. J. Botany **43**, 437—441 (1956).

HOCK, W. K.: Studies of the biodegradation and mode of antifungal action of chloroneb (1,4-dichloro-2,5-dimethoxybenzene). Ph. D. Thesis, University of Maryland, College Park, 47 p., 1968.

HOFFMAN, C., SCHWEITZER, T. R., DALBY, G.: Fungistatic properties of the fatty acids and possible biochemical significance. Food Research **4**, 539—545 (1939)

HOLZER, H.: Effect of alkylating substances on glycolysis. Chemotherapy Cancer Proc. Intern. Symp. Lugano, Switzerland, 1964, 44—50 (1964).
HORWITZ, N. H.: Methionine synthesis in Neurospora crassa. The isolation of cystathione. J. Biol. Chem. **171**, 255—264 (1947).
HORSFALL, J. G.: Fungicides and their action, pp. 1—8, 14—41. Waltham (Mass.): Chronica Botanica Company 1945.
— Principles of fungicidal action, p. 279. Waltham (Mass.): Chronica Botanica Company 1956.
— BARRATT, R. W.: An improved grading system for measuring plant diseases. Phytopathology 35—655 (1945).
— CHAPMAN, R. A., RICH, S.: Relation of structure of diphenyl compounds to fungitoxicity. Natl Res. Council Publ. **206**, 210—242 (1951).
— DIMOND, A. E.: Interaction of tissue sugar, growth substances, and disease susceptibility. Zeit. Pflanzenkrankh. u. Pflanzenschutz **64**, 415—421 (1957).
— LUKENS, R. J.: (1) Fungitoxicity of dioxanes, dioxolanes, and methylene dioxybenzenes. Conn. Agr. Expt. Sta., New Haven, Bull. **673**, 1—48 (1965).
— — (2) Fungitoxicity from irreversible water leakage. Phytopathology **55**, 1062 (1965).
— — (1) Periodic bleeding of conidiophores treated with decanoid acid. Phytopathology **56**, 882 (1966).
— — (2) Inhibiting sporulation of Alternaria solani. Abstracts of Intern. Symp. on Plant Pathology, pp. 58—59. New Delhi 1966.
— — Glycolate oxidase, a potential target for antisporulation for Alternaria solani. 6th Intern. Plant Protection Cong., pp. 319—320. Vienna 1967.
— — Aldehyde traps as antisporulants for fungi. Conn. Agr. Expt. Sta., New Haven, Bull. **694**, 1—27 (1968).
— MARSH, R. W., MARTIN, H.: Studies upon the copper fungicides. IV. The fungicidal value of the copper oxides. Ann. Appl. Biol. **24**, 867—882 (1937).
— RICH, S.: Fungitoxicity of heterocyclic nitrogen compounds. Contribs Boyce Thompson Inst. **16**, 313—347 (1951).
— — Fungitoxicity of sulfur-bridged compounds. Indian Phytopathol. **9**, 1—14 (1953).
HOTCHKISS, R. D.: Gramicidin, tyrocidin, and tyrothricin. Advances in Enzymol. **4**, 153 (1944).
HUECK, H. J., ADEMA, D. M. M., WIEGMANN, R.: Bacteriostatic, fungistatic, and algistatic activity of fatty nitrogen compounds. Appl. Microbiol. **14**, 308—319 (1966).
HUGHES, J. T.: Preliminary observations on the conversion of sodium *N*-methyldithiocarbamate (methan-sodium) to methyl isothiocyanate in soil. Rep. Glasshouse Crops Res. Inst., 1959, pp. 108—111 (1960).
IWAMATO, H., KIKUCHI, M.: Prevention of mold growth on industrial products. XIV. Mechanism of prevention of mold growth by fungicides. Kogyo Gijutsuin, Hakko Kenkyusho Kenkyu Hokoku. (Japan) **24**, 1—11 (1963).
JOHNSON, O., KROG, N., POLAND, J. L.: Pesticides, Part II: Fungicides and herbicides. Chemi. Week (June 1), 56—90 (1963).
JONES, O. T. G.: The inhibition of bacteria chlorophyll biosynthesis in Rhodopseudomonas spheroides by 8-hydroxyquinoline. Biochem. J. **88**, 335—343 (1963).
JUNIPER, B. E.: The surfaces of plants. Endeavour **18**, 20—25 (1959).
KERRIDGE, D.: The effect of actidione and other antifungal agents on nucleic acid and protein synthesis in Saccharomyces carlsbergensis. J. Gen. Microbiol. **19**, 497—506 (1958).
KIRBY, A. H. M., FRICK, E. L.: "Karathane": the relative importance of the phenolic and the unsaturated acid components in toxicity towards certain plant pathogens. Nature **182**, 1445—1446 (1958).
— — GRATWICK, M.: Greenhouse evaluation of chemicals for the control of powdery mildews. VI. Dinitro-*s*-alkyl phenols: The 'dinocap' misconception. Ann. Appl. Biol. **57**, 211—221 (1966).
KIRBY, W. M. M.: Extraction of a highly potent penicillin inactivator from penicillin resistant staphylococci. Science **99**, 452—453 (1944).
KITTLESON, A. R.: Preparation and some properties of *N*-(trichloromethylthio) tetrahydrophthalimide. J. Agr. Food Chem. **1**, 677—697 (1953).

KLÖPPING, H. L., VAN DER KERK, G. J. M.: Investigations on organic fungicides. V. Chemical constitution and fungistatic activity of aliphatic bisdithiocarbamates and isothiocyanates. Rec. trav. chim. **70**, 949—961 (1951).

KOENIG, K. H., POMMER, E. H., SANNE, W.: Novel fungicides: *N*-Substituted tetrahydro-1,3-oxazines. Angew. Chem. **77** (7), 327—333 (1965).

KOOPMANS, M. J.: An in vitro evaluation of the toxicity of chemicals for Erysiphaceae. Mededel. Landbouwhogeschool Opzoekingssta. Staat Gent. **24**, 821—827 (1959).

— Systemic fungicidal action of some 5-amino-1-bis (dimethylamido) phosphoryl triazoles-1,2,4. Mededel. Landbouwhogeschool Opzoelkingssta. Staat Gent. **25**, 1221—1226 (1960).

KORN, E. D.: Structure of biological membranes. Science **153**, 1491—1498 (1966).

KOTTKE, M., SISLER, H. D.: Effect of fungicides on permeability of yeast cells to the pyruvate ion. Phytopathology **52**, 959—961 (1962).

KUEHLE, E., KLAUKE, E., GREWE, F.: Fluorodichloromethylthio derivatives and their use in plant protection. Farbenfabriken Bayer A.-G., Angew. Chem. **76**, 807—816 (1964).

LAMPEN, J. O.: Interference by polyenic antifungal antibiotics (especially Nystatin and filipin) with specific membrane functions. Bichemical Studies of antimicrobial drugs. 16th Symposium of Soc. Gen. Microbiol., London 1966, pp. 111—130. Cambridge: Univ. Press 1966.

— ARNOW, P. M., BOROWSKA, Z., LASKIN, A. I.: Location and role of sterol at nystatin-binding sites. J. Bacteriol. **84**, 1152—1160 (1962).

— — SAFERMAN, R. S.: Mechanism of protection by sterols against polyene antibiotics. J. Bacteriol. **80**, 200—206 (1960).

LANGMUIR, I.: Overturning and anchoring of monolayers. Science **87**, 493—500 (1938).

LARGE, E. C.: The advance of the fungi, pp. 13—43. New York: Henry Holt and Co. 1940.

LAST, F. T.: The use of tetra- and penta-chloronitrobenzenes in the control of Botrytis disease and Rhizoctonia attack of lettuce. Ann. Appl. Biol. **39**, 557—568 (1952).

LEBLOVA-SVOBODOVA, S.: Effect of isothiocyanate on the synthesis of proteins. Naturwissenschaften **52**, 430 (1965).

LECLERG, E. L.: Crop losses due to plant diseases in the United States. Phytopathology **54**, 1309—1313 (1964).

LEEBRICK, J. R.: Organobismuth biocides. U.S. Patent 3,239,411 (1965).

LERMAN, L. S.: The structure of the DNA-acridine complex. Proc. Natl Acad. Sci. US **49**, 94—102 (1963).

LIEN, E. J.: Structure-activity correlation in fungitoxicity of imides and their imide-$SCCl_3$ compounds. J. Agr. Food Chem. **17**, 1265—1268 (1969).

LILLY, V. G., BARNETT, H. L.: Physiology of the fungi, 464 p. New York: McGraw-Hill Book Comp., Inc. 1951.

LINDENMAYER, A.: Carbohydrate metabolism. In: The fungi, an advance treatise. Vol. I. The fungalcide Chapter 12, pp. 301—348. (AINSWORTH, G. C., SUSSMAN, A. S., Eds) 1965.

LINEWEAVER, H., BURK, D.: The determination of enzyme dissociation constants. J. Am. Chem. Soc. **56**, 658—666 (1934).

LOCICERO, J. C., FREAR, D. E. H., MILLER, H. J.: The relation between chemical structures and fungicidal action in a series of substituted and unsubstituted pyridinium halides. J. Biol. Chem. **172**, 689—693 (1948).

LOEB, L. B.: Static electrification, pp. 25—31, 122—124. Berlin-Göttingen-Heidelberg: Springer 1958.

LOWE, M. B., PHILLIPS, J. N.: A possible mode of action of some antifungal and antibacterial chelating agents. Nature **194**, 1058—1059 (1962).

LOWERY, R. J., SUSSMAN, A. S., VON BOVENTER, B.: Physiology of the cell surface of Neurospora ascospores II. Interference with dye adsorption by polymyxin. Arch. Biochem. Biophys. **62**, 113—124 (1957).

LUCKENBAUGH, R. W.: Diphenylethylenes as fungicides. U.S. Patent 3,153,611 (1964).

LUDWIG, R. A., THORN, C. D., MILLER, D. M.: Studies on the mechanism of fungicidal action of disodium ethylene bisdithiocarbamate (nabam) Can. J. Botany **32**, 48—54 (1954).

— THORN, G. D., UNWIN, C. H.: Studies on the mechanism of fungicidal action of metallic ethylenebisdithiocarbamates. Can. J. Botany **33**, 42—59 (1955).

LUKENS, R. J.: Chemical reactions involved in the fungitoxicity of captan. Ph. D. Thesis. University of Maryland, College Park, 51 p., 1958.

— Chemical and biological studies of a reaction between captan and dialkyldithiocarbamates. Phytopathology **49**, 339—343 (1959).

— Conidial production from filter paper cultures of Helminthosporium vagans and Alternaria solani. Phytopathology **50**, 867—868 (1960).

— Studies on the chemistry of the fungicide Folcid. Phytopathology **52**, 740 (1962).

— (1) Photo-inhibition of sporulation in Alternaria solani. Am. J. Botany **50**, 720—724 (1963).

— (2) Thiophosgene split from captan by yeast. Phytopathology **54**, 881 (1963).

— Control of bluegrass foot-rot disease with a single drench of fungicide. Phytopathology **55**, 708 (1965).

— The fungitoxicity of compounds containing a trichloromethylthio group. J. Agr. Food Chem. **14**, 365—367 (1966).

— Low light intensity promotes melting out of bluegrass. Phytopathology **58**, 1058 (1968) (Abstract).

— HORSFALL, J. G.: Fungitoxicity of *N*-substituted phthalimides. Phytopathology **52**, 925 (1962).

— — Chemical constitution and fungitoxicity of imides and their imide-$SCCl_3$ compounds. Phytopathology **57**, 876—880 (1967).

— — Glycolate oxidase, a target for antisporulants. Phytopathology **58**, 1671—1676 (1968).

— — Selective inhibition of sporulation in plant pathogenic fungi. Abstracts of Papers, ACS April, 1969, 048 (1969).

— RICH, S.: Cobalt pretreatment of yeast cells increase toxicity of captan. Phytopathology **49**, 228 (1959).

— — HORSFALL, J. G.: Role of the R-group in the fungitoxicity of R-$SCCl_3$ compounds. Phytopathology **55**, 658—662 (1965).

— SISLER, H. D.: (1) Chemical reactions involved in the fungitoxicity of captan. Phytopathology **48**, 235—244 (1958).

— — (2) 2-Thiazolidinethione-4-carboxylic acid from the reaction of captan with cysteine. Science **127**, 650 (1958).

— STODDARD, E. M.: Rhizoctonia solani in relation to maintenance of golf courses. USGA J. and Turf Management **14** (4), 27—31 (1961).

LUKES, G. E., WILLIAMSON, T. B. Antimicrobial halomethyl aryl ketones. U.S. Patent 3,184,379 (1965).

LYR, H.: Enzymatische detoxifikation chlorierter phenole. Phytopathol. Z. **47**, 73—83 (1963).

MAAS, W. K., MCFALL, E.: Genetic aspects of metabolic control. Ann. Rev. Microbiol. **18**, 95—110 (1964).

MADHOSINGH, C.: The metabolic detoxication of 2,4-dinitrophenol by Fusarium oxysporium. Can. J. Microbiol. **7**, 553—567 (1961).

MANDELS, G. R.: (1) Localization of carbohydrases at the surface of fungus spores by acid treatment. Exptl Cell Research **5**, 48—55 (1953).

— (2) The properties and surface location of an enzyme oxidizing ascorbic acid in fungus spores. Arch. Biochem. Biophys. **42**, 164—173 (1953).

— Kinetics of fungal growth. In: The fungi, an advance treatise, Vol. I, Chapter 25, pp. 599—612. (AINSWORTH, G. C., SUSSMAN, S. A., Eds). New York: Academic Press 1965.

— DARBY, R. T.: A rapid cell volume assay for fungitoxicity using fungal spores. J. Bacteriol. **65**, 16—26 (1953).

MANTEN, A., KLÖPPING, H. L., VAN DER KERK, G. J. M.: Investigations on organic fungicides. II. A new method for evaluating antifungal substances in the laboratory. Antonie van Leeuwenhoek J. Microbiol. Serol. **16**, 282—294 (1950).

MARCOS, G. B., MUNICIO, A. M., VEGA, S.: Reactivity of $\alpha,\beta$-unsaturated ketones toward sulfhydryl compounds and their antifungal activity. Chem. & Ind. (London) **1964**, 2053—2054.

MARKHAM, R.: Lethal synthesis. The strategy of chemotherapy. 8th Symp. Soc. Gen. Microbiol. London, pp. 163—177. Cambridge: Univ. Press 1958.

Marrian, D. H., Friedman, E., Ward, J. L.: The antibacterial effects of substances structurally resembling maleimide. Biochem. J. **54**, 65 (1953).

Marsden, D. H.: Studies of the creosote fungus, Hormodendrum resinae. Mycologia **46**, 161—183 (1954).

Martin, H.: The scientific principles of crop protection, pp. 62—150, 5th Ed. New York: St. Martin's Press 1964.

Mason, C. L.: A study of the fungicidal action of 8-quinolinol and some of its derivatives. Phytopathology **38**, 740—751 (1948).

Mason, C. T., Brown, R. W., Minga, A. E.: The relationship between fungicidal activity and chemical constitution. Phytopathology **41**, 164—171 (1951).

Matsui, C., Nozu, M., Kilumoto, T., Matsuura, M.: Electron microscopy of fungus conidia after immersion in mercuric chloride solution. Phytopathology **52**, 88—90 (1962).

McCallan, S. E. A.: Studies on fungicides. II. Testing protective fungicides in the laboratory. Cornell Univ. Agr. Expt. Sta. Mem. **128**, 8—24 (1930).

— History of fungicides. In: Fungicides, an advanced treatise, Vol. I, pp. 1—37. (Torgeson, D. C., Ed.). New York and London: Academic Press 1967.

— Burchfield, H. P., Miller, L. P.: Interpretation of dosage-response curves. Phytopathology **49**, 544 (1959).

— Miller, L. P.: Uptake of fungitoxicants by spores. In: Perspectives of biochemical plant pathology. (Rich, S., Ed.). Conn. Agr. Exp. Sta., New Haven, Bull. **663**, 137—148 (1963).

— — Weed, R. M.: Comparative effects of fungicides on oxygen uptake and germination of spores. Contribs Boyce Thompson Inst. **18**, 39—68 (1954).

— Wellman, R. I., Wilcoxon, F.: An analysis of factors causing variation in spore germination tests of fungicides. III. Slope of toxicity curves, replicate tests, and fungi. Contribs Boyce Thompson Inst. **12**, 49—78 (1941).

— Wilcoxon, F.: Fungicidal action and the periodic system of the elements. Contribs Boyce Thompson Inst. **6**, 479—500 (1934).

— — The action of fungus spores on Bordeaux mixture. Contribs Boyce Thompson Inst. **8**, 151—165 (1936).

— — An analysis of factors causing variation in spore germination tests of fungicides. Phytopathology **29**, 16 (1939).

McNew, G. L., Burchfield, H. P.: Fungitoxicity and biological activity of quinones. Contribs Boyce Thompson Inst. **16**, 357—374 (1951).

— Sundholm, N. K.: The fungicidal activity of substituted pyrazoles and related compounds. Phytopathology **39**, 721—751 (1949).

Mellor, D. P., Maley, L.: Order of stability of metal complexes. Nature **161**, 436—437 (1948).

Menzies, J. D.: Dosage rates and application methods with PCNB for control of potato scab and Rhizoctonia. Am. Potato J. **34**, 219—226 (1957).

Merker, R. L.: Some silyacetylenes as bacteriocides and fungicides. French Patent 1,403,705 (1965).

Meyer, A.: The utilization of organic coloring materials and 8-hydroxyquinoline in combating cryptogamic disease of the vine. Rev. viticult. **77**, 117—120 (1932).

Miller, H. J.: Relationship of concentration of some organic substances to spore germination and dosage response. Phytopathology **40**, 326—332. (1950).

Miller, L. P.: Natural membranes. In: Diffusion and membrane technology. Chapter 17, pp. 345—367. (Tuwiner, S. B., Ed.). New York: Reinhold Publishing 1962.

— McCallan, S. E. A.: Toxic action of metal ions to fungus spores. J. Agr. Food Chem. **5**, 116—122 (1957).

— — Weed, R. M.: Quantitative studies on the role of hydrogen sulfide formation in the toxic action of sulfur to fungus spores. Contribs Boyce Thompson Inst. **17**, 151—171 (1953).

— Richter, E.: Relationship of lipid contents of fungal conidia to uptake of toxicants. Phytopathology **50**, 646 (1960).

Miller, P. M.: Heat-decomposition products of captan as phytotoxic agents. Phytopathology **47**, 245 (1957).

— LUKENS, R. J.: Deactivation of sodium *N*-methyldithiocarbamate in soil by nematicides containing halogenated hydrocarbons. Phytopathology **56**, 967—970 (1966).
— STODDARD, E. M.: Importance of fungicide volatility in controlling soil fungi. Phytopathology **47**, 24 (1957).
MISATO, T.: Mode of action of agricultural antibiotics developed in Japan. Experimental approaches to pesticide metabolism, degradation and mode of action. U.S.-Japan Seminar, Nikko, Japan, Aug. 16—19, 1967, pp. 83—92.
MITCHELL, J. W.: Progress in research on absorption, translocation, and exudation of biologically active compounds in plants. In: Prospectives of biochemical plant pathology. (RICH, S., Ed.). Conn. Agr. Exp. Sta., New Haven, Bull. **663**, 49—56 (1963).
— BROWN, J. W.: Movement of 2,4-dichlorophenoxyacetic stimulus and its relation to the translocation of organic food materials in plants. Botan. Gaz. **107**, 393—407 (1946).
MOJE, W.: The chemistry and nematocidal activity of organic halides. In: Advances in Pest Control Research. (METCALF, R. L., Ed.). **3**, 181—217 (1960).
— KENDRICK, J. B., JR., ZENTMYER, G. A.: Systemic and fungicidal activity of D-, L-, homocysteine derivatives, and methionine antagonists. Phytopathology **53**, 883 (1963).
MOMOKI, K., UENO, A., AKAGI, S., OBATA, M., SAKAI, S.: Antitrichomonas activity of 5-nitrofuram derivatives. Nippon Kagaku Rydhogakukai Zasshi **12** (5), 371—374 (1964).
MONTEITH, J., DAHL, A. S.: Turf diseases and their control. Bull. U.S. Golf Assoc., Green Sect. **12** (4) 85—186 (1932).
MONTIE, T. C., SISLER, H. D.: Effects of captan on glucose metabolism and growth of Saccharomyces pastorianus. Phytopathology **62**, 94—102 (1962).
MOREHART, A. L., CROSSAN, D. F.: Studies on the ethylene bisdithiocarbamate fungicides. Delaware, Univ., Agr. Expt Stat., Bull. **357**, 26 p. (1965).
MORTIMER, R. K., JOHNSTON, J. R.: Life span of individual yeast cells. Nature **183**, 1751 to 1752 (1959).
MOYED, H. S.: Interference with the feed-back control of histidine biosynthesis. J. Biol. Chem. **236**, 2261—2267 (1961).
MÜLLER, E., BIEDERMANN, W.: Der Einfluß von $Cu^{+2}$ Ionen auf den Keimungablauf von Alternaria tenuis. Phytopathol. Z. **19**, 343—350 (1952).
MUNNECKE, D. E., DOMSCH, K. H., ECKERT, J. W.: Fungicidal activity of air passed through columns of soil treated with fungicides. Phytopathology **52**, 1298—1306 (1962).
— MARTIN, J. P.: Release of methylisothiocyanate from soils treated with Mylone (3,5-dimethyl-tetrahydro-1,3,5,2-H-thiadiazine-2-thione). Phytopathology **54**, 941—945 (1964).
NELSON, K. E., RICHARDSON, H. B.: Storage temperature and sulfur dioxide treatment in relation to decay and bleaching of stored table grapes. Phytopathology **57**, 950—959 (1967).
NEMEC, P., DROBNICA, L., ANTOS, K., KRISTIAN, P., HULKA, A.: Interrelations between the structure and biological activity of some isothiocyanates. Nat. Occurring Goitrogens Thyroid Function Symp. Smolenice, Czek. 1962, 71—78 (1964).
NEWHALL, A. G., LEAR, B.: Soil fumigation for nematode and disease control. Cornell Univ., Agr. Expt Sta. Bull. **859**, 1—32 (1948).
NEWTON, B. A.: Surface-active bacteriocides. In: The strategy of chemotherapy. Symposium Soc. Gen. Microbiol. **8**, 62—93 (1958).
NICKELL, L. G., GORDON, P. N., GOENAGA, A.: 2-*n*-Alkylmercapto-1,4,5,6-tetrahydropyrimidines, chemotherapeutic agents for plant rusts. Plant Disease Reptr **45**, 756—758 (1961).
NICKERSON, W. J., VAN RIJ, N. J. W.: The effect of sulfhydryl compounds, penicillin, and cobalt on the cell division mechanism of yeast. Biochim. et Biophys. Acta **3**, 461—475 (1949).
NOVIKOV, S. S., SOLOV'EV, V. N., OVCHEROV, K. E., KHMEL'NITSKII, L. I., NOVIKOVA, T. S., KOYAEV, G. A., BORISOVA, N. N.: Physiological activity of simple nitrofuron derivatives. Nitro Compounds, Proc. Intern. Symp., Warsaw 1963, 519—531 (1963).
OKU, H., NAKANISHI, T.: Mode of action of an antibiotic, asochytin, with reference to selective toxicity. Phytopathol. Z. **55**, 1—14 (1966).
ORDISH, G., MITCHELL, J. F.: World fungicide usage. In: Fungicides, an advanced treatise, Vol. I, pp. 39—62. (TORGESON, D. C., Ed.). New York and London: Academic Press 1967.

OSTER, K. A., WOODSIDE, R.: Fungistatic and fungicidal compounds. In: Disinfection, sterilization, and preservation, pp. 305—320. (LAWRENCE, C. A., BLOCK, S. S., Eds). Philadelphia: Lea and Febiger 1968.

OSTER, R.: Results of irradiating Saccharomyces with monochromatic ultraviolet light. III. The influence of modifying factors. J. Gen. Physiol. **18**, 243—250 (1934).

OWENS, R. G.: (1) Studies on the nature of fungicidal action. I. Inhibition of sulfhydryl, amino, iron, and copper-dependent enzymes in vitro by fungicides and related compounds. Contribs Boyce Thompson Inst. **17**, 221—242 (1953).

— (2) Studies on the nature of fungicidal action. II. Chemical constitution of benzenoid and quinonoid compounds in relation to fungitoxicity and inhibition of amino- and sulfhydryl-dependent enzymes. Contribs Boyce Thompson Inst. **17**, 273—282 (1953).

— Studies on the nature of fungicidal action. III. Effects of fungicides on polyphenol oxidase in vitro. Contribs Boyce Thompson Inst. **17**, 473—487 (1954).

— Effects of elemental sulfur, dithiocarbamates, and related fungicides on organic acid metabolism of fungus spores. Developments in Industrial Microbiology **1**, 187—205 (1960).

— Chemistry and physiology of fungicidal action. Ann. Rev. Phytopathology, Ann. Rev. Inc. Palo Alto, California **1**, 77—100 (1963).

— BLAAK, B.: (1) Site of action of captan and dichlone in the pathway between acetate and citrate in fungus spores. Contribs Boyce Thompson Inst. **20**, 459—474 (1960).

— — (2) Chemistry of the reactions of dichlone and captan with thiols. Contribs Boyce Thompson Inst. **20**, 475—497 (1960).

— HAYES, A. D.: Biochemical action of thiram and some dialkyl dithiocarbamates. Contribs Boyce Thompson Inst. **22**, 227—240 (1964).

— MILLER, L. P.: Intracellular distribution of metal ions and organic fungicides in fungus spores. Contribs Boyce Thompson Inst. **19**, 177—188 (1957).

— NOVOTNY, H. M.: Mechanism of action of the fungicide dichlone (2,3-dichloro-1,4-naphthoquinone). Contribs Boyce Thompson Inst. **19**, 463—482 (1958).

— — Mechanism of action of the fungicide captan (*N*-trichloromethylthio)-4-cyclohexene-1,2-dicarboximide). Contribs Boyce Thompson Inst. **20**, 171—190 (1959).

— RUBINSTEIN, J. H.: Chemistry of the fungicidal action of tetramethylthiuram disulfide (thiram) and ferbam. Contribs Boyce Thompson Inst. **22**, 241—257 (1964).

PAGE, G. E.: Soil fumigation equipment. Oregon State Univ. Expt Bull. **813**, 1—14 (1963).

PASSOW, H., ROTHSTEIN, A.: The binding of mercury by the yeast cell in relation to changes in permeability. J. Gen. Physiol. **43**, 621—633 (1960).

PATIL, S. S., DIMOND, A. E.: (1) Repression of polygalacturonase synthesis in Fusarium oxysporium f. s. lycopersici by sugars and its effect on symptom reduction in infected tomato plants. Phytopathology **58**, 676—682 (1968).

— — (2) Effect of phenols and cytokinins on polygalacturonase content of Verticillium cultures. Phytopathology **58**, 868—869 (1968).

PAUL, J.: Cell Biology, pp. 77—91. Stamford, California: Stamford University Press 1964.

PELLEGRINI, G., BUGIANI, A., TENERINI, I.: Systemic properties of a new class of fungicides Phytophatol. Z. **52**, 37—48 (1965).

PERMOGOROV, V. I., PROZOROV, A. A., SHEMYAKIN, M. F., LAZURKIN, YU. S., KHESIN, R. B.: A mechanism of the inhibition of the biological activity of DNA by actinomycin. Mol. Akad. Nauk SSSR, Inst. Biol. Fiz. Sb. Statei **1965**, 162—180.

PIANKA, M., POLTON, D. J.: Substituted phenyl carbonates, thiolo carbonates and thionothiolo carbonates as acaricides, ovicides, fungicides, and insecticides. U.S. Patent 3,234,260, Feb. 8, 1966.

POTTS, S. F.: Ground spray equipment and aerial equipment. In: Concentrated spray equipment, mixtures and application methods, pp. 115—358. New Jersey: Dorland Books, Caldwell 1958.

PRÉVOST, B.: Mémoire sur la cause immédiate de la carie ou charbon des blés et de plusieurs autres maladies des plantes, et sur les préservatifs de la carie. Phytopathology Classic **6**, 1—4. Trans. by KEITT, G. W. (1939) (1807).

PRIEST, D., WOOD, R. K. S.: Strains of Botrytis allei resistant to chlorinated nitrobenzenes. Ann. Appl. Biol. **49**, 445—460 (1961).

PTITSYNA, N. V., DURDINA, O. A.: Copper naphthenate paste as substitute for Bordeaux mixture. Zashchita Rastenii ot Vreditelei i Boleznei **6** (3), 36—38 (1961).

PURDY, L. H.: Application and use of soil- and seed-treatment fungicides. In: Fungicides, an advance treatise. Vol. I, Chapter 7, pp. 195—237 (TORGESON, D. C., Ed.). New York City: Academic Press 1967.

QUASTEL, J. M.: Membrane structure and function. Science **158**, 146—161 (1967).

RABANUS, A.: Über die Säure-Produktion von Pilzen und deren Einfluß auf die Wirkung von Holzschutzmitteln. Mitt. deut. Forstver. **23**, 77—89 (1939).

RADAR, W. E., MONROE, C. M., WHETSTONE, R. R.: Tetrahydropyrimidine derivatives as potential foliage fungicides. Science **115**, 124—125 (1952).

RAMSEY, G. B., SMITH, M. A., HEIBERG, B. C.: Fungistatic action of diphenyl on citrus fruit pathogens. Botan. Gaz. **106**, 74—83 (1944).

RAZIN, S., MOROWITZ, M. J., TERRY, T. M.: Membrane subunits of Mycoplasma laidlawii and their assembly to membrane structure. Natl Acad. Sci. US **54**, 219—225 (1965).

REYNOLDS, P. E.: Antibiotics affecting cell-wall synthesis. Biochemical studies of antimicrobial drugs. 16th Symposium of Soc. Gen. Microbiol., pp. 47—69, London 1966. Cambridge: Univ. Press 1966.

RICE, E. L., ROHRBAUGH, L. M.: Effect of kerosene on movement of 2,4-dichlorophenoxy acetic acid and some derivatives through destarched bean plants in darkness. Botan. Gaz. **115**, 76—81 (1953).

RICH, S.: Dynamics of deposition and tenacity of fungicides. Phytopathology **44**, 203—213 (1954).

— Foliage fungicides plus glycerin for the chemotherapy of cucumber scab. Plant Disease Reptr **40**, 620—621 (1956).

— Fungicidal chemistry. In: Plant pathology, an advance treatise, Vol. II, pp. 553—602 (HORSFALL, J. G., DIMOND, A. E., Eds). New York: Academic Press 1960.

— Quinones. In: Fungicides, an advance treatise, pp. 447—475 (TORGESON, D. C., Ed.). New York: Academic Press 1968.

— HORSFALL, J. G.: Metal reagents as antisporulants. Phytopathology **38**, 22 (1948).

— — Fungicidal activity of dinitrocaprylphenyl crotonate. Phytopathology **39**, 19 (1949).

— — The relation between fungitoxicity, permeation, and lipid solubility. Phytopathology **42**, 457—460 (1952).

— — (1) Relation of polyphenol oxidase to fungitoxicity. Proc. Natl Acad. Sci. US **40**, 139—145 (1954).

— — (2) Fungitoxicity of ethylenethiourea derivatives. Science **120**, 122—123 (1954).

— — KEIL, H. L.: The relation of laboratory to field performance of fungicides. Rept Int. Hort. Congress **13**, 282—287 (1953).

RICHARDSON, L. T.: Reversal of fungitoxicity of thiram by seed and root exudates. Can. J. Botany **44**, 111—112 (1966).

— THORN, G. D.: The interaction of thiram and spores of Glomerella cingulata Spauld. and Schrenk. Can. J. Botany **39**, 531—540 (1961).

RICHMOND, D. V., SOMERS, E.: Studies of the fungitoxicity of captan. II. The uptake of captan by conidia of *Neurospora crassa*. Ann. Appl. Biol. **50**, 45—56 (1962).

— — Studies on the fungitoxicity of captan. III. Relation between sulfhydryl content of fungal spores and their uptake of captan. Ann. Appl. Biol. **52**, 327—336 (1963).

— — Studies on the fungitoxicity of captan. IV. Reactions of captan with cell thiols. Ann. Appl. Biol. **57**, 231—240 (1966).

— — MILLINGTON, P. F.: Studies on the fungitoxicity of captan. V. Electron microscopy of captan-treated Neurospora crassa conidia. Ann. Appl. Biol. **59**, 233—237 (1967).

RICHMOND, M. H.: Structural analogy and chemical reactivity in the actions of antibacterial compounds. In: Biochemical studies of antimicrobial drugs. 16th Symp. Soc. Gen. Microbiol., pp. 301—334, London 1966. Cambridge: University Press 1966.

RIGLER, N. E., GREATHOUSE, G. A.: Fungicidal potency of quinoline homologues and derivatives. Ind. and Eng. Chem. **33**, 693—694 (1941).

RIPPER, W. E.: Application methods for crop protection chemicals. Ann. Appl. Biol. **42**, 288—324 (1955).

ROBERTSON, J. D.: The ultrastructure of cell membranes and their derivatives. Biochem. Soc. Symp. (Cambridge, England). **16**, 3—43 (1959).

ROTHSTEIN, A., HAYES, A. D.: The relationship of the cell surface to metabolism. XIII. The cation-binding properties of the yeast cell surface. Arch. Biochem. Biophys. **63**, 77—79 (1956).

SALTON, M. R. J.: Structure and function of bacterial cell membranes. Ann. Rev. Microbiol. **21**, 417—442 (1967).

SANDERS, S. L., NELSON, C. T.: Treatment of the superficial mycoses. In: Fungi and fungous diseases, pp. 307—310. (DALLDORF, G., Ed.). Springfield, Illinois: Charles C. Thomas 1962.

SCHMITT, C. G.: Comparison of a series of derivatives of 4,5-dimethyl-2-mercaptothiazole for fungicidal efficacy. Contribs Boyce Thompson Inst. **16**, 261—265 (1951).

SCHOOT, C. J., KOOPMANS, M. J., VAN DER BOS, B. G.: Bis(dimethylamido) pentachlorophenol fungicidal compositions. U.S. Patent 3,157,568 (1964).

SCHULDT, P. H., WOLF, C. N.: Fungitoxicity of substituted *s*-triazines. Contribs Boyce Thompson Inst. **18**, 377—393 (1956).

SEMPIO, C.: Sulla interpretazione del meccanismo intimo di azione dello zolfo come anticrittogamico. Mem. reale accad. Italia, Classe sci. fis. mat. e nat. **3**, Biol. **2**, 1—30 (1932).

SEXTON, W. A.: Chemical constitution and biological activity, pp. 260—275. 2. ed. New York, Toronto, London: D. von Nostrand Co., Inc. 1953.

SHANNON, E. L., CLARK, W. S., REINHOLD, G. W.: Chlorine resistance of enterococci. J. Milk and Food Technol. **28**, 120—123 (1965).

SHAW, E., BERNSTEIN, J., LOSEE, K., LOTT, W. A.: Analogs of aspergillic acid. IV. Substituted 2-bromo pyridine-*N*-oxides and their conversion to cyclic thiohydroxamic acids. J. Am. Chem. Soc. **72**, 4362—4364 (1950).

SHAW, W. H. R.: Toxicity of cations toward living systems. Science **120**, 361—363 (1954).

SHEPARD, C. J., MANDRYK, M.: Effects of metabolites and antimetabolites on the sporulation of Peronospora tabacina Adam. on tobacco leaf disks. Australian J. Biol. Sci. **17**, 878 to 891 (1964).

SHEPARD, E. R., SHONLE, H. A.: Imidazolinium salts as topical antiseptics. J. Am. Chem. Soc. **69**, 2269—2270 (1947).

SHEPARD, H. H., MAHAN, J. N., FOWLER, D. L.: The Pesticide Review 1966, 32 p. Ag. Stabilization and Conservation Service. Washington, D. C.: U.S.D.A.1966.

SHEPARD, M. C.: Control of cucurbit powdery mildews with a new systemic fungicide. Abstracts of First Int'l Cong. Plant Pathology (London), 180 (1968).

SHIOYAMA, OSAMU, HURONO, HITOSHI, MURATA, KIKUGO, MATSUMOTO, SEIZO: Fungicidal effect of organoarsenic compounds against rice stem rot fungus. Noyaku Seisan Gijutsu **11**, 8—12 (1964).

SHIVE, W., SKINNER, C. G.: Amino acid analogues. In: Metabolic inhibitors, Vol. 1, pp. 1—73 (HOCHESTER, R. M., QUASTEL, J. H., Eds). New York: Academic Press 1963.

SHOMOVA, E. A., RUDAVSKII, V. P., KHASKIN, I. G.: Fungicidal activity of some aromatic trichloroacetamide derivatives. Mikrobiologiya **34**, 715—719 (1965).

SHURTLEFF, M. C., TAYLOR, D. P., COURTER, J. W., PETTY, H. B.: Soil disinfection — methods and materials. Illinois, Univ., Coll. Agr. Ext. Serv., Circ. **893**, 1—23 (1957).

SIEGEL, M. R., SISLER, H. D.: Inhibition of protein synthesis *in vitro* by cycloheximide. Nature **200**, 675—676 (1963).

— — Site of action of cycloheximide in cells of Saccharomyces pastorianus. II. The nature of inhibition of protein synthesis in a cell-free system. Biochim. et Biophys. Acta **87**, 83—89 (1964).

— — Site of action of cycloheximide in cells of Saccharomyces pastorianus. III. Further studies on the mechanism of action and the mechanism of resistance in Saccharomyces species. Biochim. et Biophys. Acta **103**, 558—567 (1965).

— — Metabolic fate of $C^{14}$ and $S^{35}$ labelled phaltan (*N*-trichloromethylthiophthalimide) in cells of Saccharomyces pastorianus. Phytopathology **57**, 831 (1967).

— — (1) Fate of the phthalimide and trichloromethylthio ($SCCl_3$) moieties of folpet in toxic action on cells of Saccharomyces pastorianus. Phytopathology **58**, 1123—1128 (1968).

— — (2) Reaction of folpet with purified enzymes, nucleic acids and subcellular components of Saccharomyces pastorianus. Phytopathology **58**, 1129—1133 (1968).

— — Johnson, F.: Relationship of structure to fungitoxicity of cycloheximide and related glutarimide derivatives. Biochem. Pharmacol. **15**, 1213—1223 (1966).

Sijpesteijn, A. K., Dekhuijzen, H. M., Kaslander, J., Pluijgers, C. W., van der Kerk, G. J. M.: Metabolism of sodium dimethyldithiocarbamate by plants and microorganisms. Mededeel. Landbouwhogeschool Opzoekingssta. Staat Gent **28**, 597—603 (1963).

— Janssen, M. J.: Fungitoxic action of 8-hydroxyquinoline, pyridine-*N*-oxide-2-thiol and sodium dialkyldithiocarbamates, and their copper complexes. Nature **182**, 1313—1314 (1958).

— — Dekhuijzen, H. M.: Effect of copper and chelating agents on growth inhibition of Aspergillis niger by 8-hydroxyquinoline and pyridine-*N*-oxide-2-thiol. Nature **180**, 505—506 (1957).

— Pluijers, C. W.: Phenylthioureas as systemic fungicides. Mededeel. Landbouwhogeschool Opzoekingssta. Staat Gent **27**, 1199—1203 (1962).

— van der Kerk, G. J. M.: Fate of fungicides in plants. Ann. Rev. Phytopathology. Ann. Rev. Inc. Palo Alto **3**, 127—152 (1965).

Sisler, H. D.: Fungitoxic mechanisms. In: Perspectives of biochemical plant pathology (Rich, S., Ed.). Conn. Agr. Exp. Sta, New Haven, Bull. **663**, 116—133 (1963).

— Siegel, M. R.: Cycloheximide and other glutarimide antibiotics. In: Antibiotics, Vol. I, pp. 283—307 (Gottlieb, D., Shaw, P., Eds). Berlin-Heidelberg-New York: Springer 1967.

— — Ragsdale, Nancy: Factors regulating toxicity of cycloheximide derivatives. Phytopathology **57**, 1191—1196 (1967).

Smale, B. C.: Effects of certain trace metals on the fungitoxicity of sodium dimethyldithiocarbamate. Ph. D. Thesis. University of Maryland, College Park, Maryland, U. S. A. p. 44, 1957.

Somers, E.: Studies of spray deposits. III. Factors influencing the level of "run-off" deposits of copper fungicides. J. Sci. Food Agr. **8**, 520—526 (1957).

— The fungitoxicity of metal ions. Ann. Appl. Biol. **49**, 246—253 (1961).

— (1) The uptake of copper by fungal cells. Ann. Appl. Biol. **51**, 525—537 (1963).

— (2) The uptake of dodine acetate by Neurospora crassa. Overdruk uit de Mededelingen van de Landbouwhogeschool en de Opzoekingstations van de Staat de Gent, 1963. Deel XXVIII (3) 580—588 (1963).

— The sites of reaction of fungicides in spores. Proc. 18th Symposium Colston Research Society, Univ. of Bristol 1966, **18**, 299—308 (1967).

— Pring, R. J.: Uptake and binding of dodine acetate by fungal spores. Ann. Appl. Biol. **58**, 457—466 (1966).

— Richmond, D. V., Pickard, J. A.: Carbonyl sulphide from the decomposition of captan. Nature **215**, 214 (1967).

Sosnovsky, C.: Trichloromethylthio-derivatives of biological interest. J. Chem. Soc. **1956**, 3139—3141.

Spanis, W. C., Munnecke, D. E., Solberg, R. A.: Biological breakdown of two organic mercurial fungicides. Phytopathology **52**, 455—462 (1962).

Spencer, E. Y.: Structure and activity relation among fungicides. In: Perspectives of biochemical plant pathology (Rich, S., Ed.). Conn. Agr. Exp. Sta., New Haven, Bull. **663**, 95—112 (1963).

Spencer, H., Rosoff, B.: Effect of chelating agents on the removal of zinc 65 in man. Health Phys. **12** (4), 475—480 (1966).

Stadtman, E. R.: Allosteric regulation of enzyme activity. Advances in Enzymol. **28**, 41—154 (1966).

Stark, F. L.: Investigations on chloropicrin as a soil fumigant. Cornell Agr. Exp. Sta. Mem. **278**, 1—61 (1948).

Sternberg, T. H., Newcomer, V. D.: Therapy of fungus diseases, 337 p. Boston: Little, Brown, and Co. 1955.

Sussman, A. S.: In: Perspectives of biochemical plant pathology (Rich, S., Ed.). Conn. Agr. Exp. Sta., New Haven, Bull. **663**, 148—150 (1963).

— Lowery, R. J.: Physiology of the cell surface of Neurospora ascopores. I. Cation binding properties of the cell surface. J. Bacteriol. **70**, 675—685 (1955).
— von Boventer-Heidenhain, B., Lowery, R. J.: Physiology of the cell surface of Neurospora ascospores. IV. The functions of surface binding sites. Plant Physiol. **32**, 586—590 (1957).
Takebayashi, M., Shingaki, T., Mihara, T.: Reaction of thiols with propylene oxide in the presence of free radical sources. Bull. Chem. Soc., Japan **39**, 376—379 (1966).
Thatcher, R. W., Streeter, L. R.: The adherence to foliage of sulfur in fungicidal dusts and sprays. N.Y. State Agr. Exp. Sta. (Geneva, N. Y.), Bull. **116**, 1—18 (1925).
Thorn, G. D., Ludwig, R. A.: *S*-Alkyl-2,5-dimercapto-1,3,4-thiadiazoles. Can. J. Botany **36**, 389—392 (1958).
— — The dithiocarbamates and related compounds, pp. 169—185, 224—271. Amsterdam-New York: Elsevier 1962.
— Richardson, L. T.: Ferbam — some observations. Mededeel. Landbouwhogeschool Opzoekingssta. Staat. Gent. **27**, 1175—1178 (1962).
— — Decomposition of ferbam. Phytopathology **54**, 910 (1964).
Thornberry, H. H.: A paper disk plate method for the quantitative evaluation of fungicides and bactericides. Phytopathology **40**, 419—429 (1950).
Tisdale, W. H., Flenner, A. L.: Derivatives of dithiocarbamic acid as pesticides. Ind. and Eng. Chem. **34**, 501—502 (1942).
Tolkmith, H., Seiber, J. N., Budde, P. E., Mussell, D. R.: Imidazole: fungitoxic derivatives. Science **158**, 1462—1463 (1967).
Tolmsoff, W. J.: Biochemical basis for biological specificity of Dexon (*p*-dimethylaminobenzenediazo sodium sulfonate) as a fungistat. Phytopathology **52**, 755 (1962).
Torgeson, D. C.: Fungicides, an advanced treatise. Vol. I, 697 p. Agricultural and industrial applications; environmental interactions. New York and London: Academic Press 1967.
— Fungicides, an advanced treatise. Vol. II, 742 p. Chemistry and physiology. New York and London: Academic Press 1969.
— Hensley, W. H., Lambrech, J. A.: *N*-phenylmaleimides and related compounds as soil fungicides. Contribs Boyce Thompson Inst. **22**, 67—71 (1963).
— Yoder, D. M., Johnson, J. B.: Biological activity of Mylone breakdown products. Phytopathology **47**, 536 (1957).
Turner, N. J., Limpel, L. E., Battershell, R. D., Bluestone, H., Lamont, D.: A new foliage protectant fungicide tetrachloroisophthalonitrile. Contribs Boyce Thompson Inst. **22**, 303—310 (1964).
Tweedy, B. G.: A possible mechanism for the reduction of elemental sulfur by Monilinia fructiola. Phytopathology **54**, 910 (1964).
Utz, J. P.: Clinical application and side effects of antifungal agents. In: Systemic mycoses pp. 242—252 (Wolstenholme, G. E. W., Porter, Ruth, Eds.). Little, Brown and Co. 1967.
Vaartaja, O.: Chemical treatment of seedbeds to control nursery disease. Botan. Rev. **30**, 1—91 (1964).
Van der Kerk, G. J. M.: The present state of fungicide research. Achtste Jarlijks Symposium over Phytopharmacie. Mededeel. Landbouwhogeschool Gent **21**, 305—339 (1956).
— Chemical structure and fungicidal activity of dithiocarbamates. In: Plant pathology, problems and progress 1908—1958, pp. 280—290 (Holton, G. S., Ed.). Madison: Univ. Wisconsin Press 1959.
— Fungicides, retrospect and prospect. World Rev. Pest Control. **2** (Part 3), 29—41 (1963).
— Klöpping, H. L.: Investigations on organic fungicides. VII. Further considerations regarding the relation between chemical structure and antifungal action of dithiocarbamates and bisdithiocarbamate derivatives. Rec. trav. chim. **71**, 1179—1197 (1952).
Van Overbeek, J.: Absorption and translocation of plant regulators. Ann. Rev. Plant Physiol. **7**, 355—372 (1956).
Van Zutphen, H., van Dennen, L. L. M., Kinsky, S. C.: Action of polyene antibiotics on bilayer lipid membranes. Biochem. Biophys. Res. Commun. **22**, 393—398 (1966).
Vazquez, D.: Mode of action of chloramphenical and related antibiotics. In: Biochemical studies of antimicrobial drugs, pp. 169—191. 16th Symp. Soc. Gen. Micribiol., London 1966. Cambridge: Univ. Press 1966.

VOGEL, H. J.: Repressed and induced enzyme formation: a unified hypothesis. Proc. Natl Acad. Sci. US **43**, 491—496 (1957).
VON SCHMELING, B.: Fungicidal activity of *N*-aryl itaconimides. Phytopathology **52**, 819 to 822 (1962).
— KULKA, M.: Systemic fungicidal activity of 1,4-oxathiin derivatives. Science **152**, 659 to 660 (1966).
WADSWORTH, D. F.: Chemical control of disease affecting turf on golf greens. Golf Course Reptr **28** (3), 26—30 (1960).
WAGNER, A., BECK, W., DISKUS, A., ZELLNER, F.: Fungicides. Austrian Patent 237, 380 (1964).
WARING, M. J.: Cross-linking and intercalation in nucleic acids. Biochemical studies of antimicrobial drugs, pp. 235—265. 16th Symp. Soc. Gen. Microbiol., London 1966. Cambridge: Univ. Press 1966.
WEBER, D. J., OGAWA, J. M.: The mode of action of 2,6-dichloro-4-nitroaniline in Rhizopus arrhizus. Phytopathology **55**, 159—165 (1965).
WECK, F. J., STERN, D. R.: Ortho-substituted phenol fungicides. French patent 1,356,495 (1964).
WEDDING, R. T., KENDRICK, J. B., JR.: Toxicity of *N*-methyl dithiocarbamate and methyl isothiocyanate to Rhizoctonia solani. Phytopathology **49**, 557—561 (1959).
WEIBULL, W.: A statistical distribution function of wide applicability. J. Appl. Mechanics **18**, 293—297 (1951).
WEIDNER, J. P., BLOCK, S. S.: Alkyl and aryl thiosulfonates. J. Med. Chem. **7**, 671 (1964).
WEIL, E. D., GEERING, E. J., SMITH, K. J.: Tetrahaloethylsulfenyl halides as nematocides. U.S. Patent 3,156,611 (1964).
WEINBACH, E. C., GARBUS, J.: Protein as the mitochondrial site for action of uncoupling phenols. Science **145**, 824—826 (1964).
WELLMAN, R. H., MCCALLAN, S. E. A.: Correlations within and between laboratory slide-germination, greenhouse tomato foliage disease, and wheat smut methods of testing fungicides. Contribs Boyce Thompson Inst. **13**, 143—169 (1943).
— — Glyoxalidine derivatives as foliage fungicides. I. Laboratory studies. Contribs Boyce Thompson Inst. **14**, 151—160 (1946).
WESCOTT, E. W., SISLER, H. D.: Uptake of cycloheximide by a sensitive and a resistant yeast. Phytopathology **54**, 1261—1264 (1964).
WESSEL, C. J., BEJUKI, W. M.: Industrial fungicides. Ind. and Eng. Chem. **51**, 62A—63A (1959).
WEST, B., WOLF, F. T.: The mechanism of action of the fungicide, 2-heptadecyl-2-imidazoline. J. Gen. Microbiol. **12**, 396—401 (1955).
WETTSTEIN, F. O., NOLL, H., PENMAN, S.: Effect of cycloheximide on ribosomal aggregates engaged in protein synthesis in vitro. Biochim. et Biophys. Acta **87**, 525—527 (1964).
WILCOXON, F., MCCALLAN, S. E. A.: The fungicidal action of sulphur, III. Physical factors affecting the efficiency of dusts. Contribs Boyce Thompson Inst. **3**, 509—528 (1931).
— — Fungicidal action of organic thiocyanates, resorcinol derivatives, and other organic compounds. Contribs Boyce Thompson Inst. **7**, 333—340 (1935).
— — Theoretical principles underlying laboratory toxicity tests of fungicides. Contribs Boyce Thompson Inst. **10**, 329—338 (1939).
WILSON, H. F., JANES, R. J., CAMPAU, E. J.: Electrostatic charge effects produced by insecticidal dusts. J. Econ. Entomol. **37**, 651—655 (1944).
WOODCOCK, D.: Microbial detoxication and other transformations. In: Fungicides, an advanced treatise, Vol. I, pp. 613—642 (TORGESON, D. C., Ed.). New York: Academic Press 1967.
— BYRDE, R. J. W.: Antifungal naphthalene derivatives, with special reference to apple mildew. Mededel. Landbouwhogeschool Opzoekingssta. Staat Gent. **28**, 568—573 (1963).
YARWOOD, C. E. Fungicides for powdery mildews. Proc. Second. Internat. Congress of Crop Protection, pp. 1—22, 1951.
YODER, D. M.: Reversibility of copper toxicity to conidia of Sclerotinia fructicola. Phytopathology **41**, 39 (1951).

YURA, T., VOGEL, H. J.: Pyrroline-5-carboxylate reduction of Neurospora crassa: partial purification and some properties. J. Biol. Chem. **234**, 335—338 (1959).
ZALOKAR, M.: Integration of cellular metabolism. In: The fungi, an advance treatise, Vol. I, Chapter 14, pp. 337—426 (AINSWORTH, G. C., SUSSMAN, A. S., Eds). The fungal cell 1965.
ZARACOVITIS, C.: Factors in testing fungicides against powdery mildews. I. A slide spore-germination method of evaluating protectant fungicides. Ann. Appl. Biol. **54**, 361—374 (1964).
ZAYED, S. M. A. D., MOSTAFA, I. Y., FARGHALY, M.: Preparation and fungicidal properties of some arylthioalkyl-aryl and arylsulfonylacetohydroxamic acids. Z. Naturforsch. **21** (2) b, 180—182 (1966).
ZEDLER, R. J.: Biocides. Brit. patent 1,022,025 (1966).
ZELITCH, I.: Control of leaf stomata — their role in transpiration and photosynthesis. Am. Scientist **55**, 472—486 (1967).
ZENTMYER, G. A.: Inhibition of metal catalysis as a fungistatic mechanism. Science **100**, 294—295 (1944).
— Mechanism of action of 8-hydroxyquinoline. Phytopathology **33**, 1121 (1943).
— RICH, S.: Reversal of fungitoxicity of 8-quinolinol and copper-8-quinolinolate by other chelators. Phytopathology **46**, 33 (1956).
— — HORSFALL, J. G.: Reversal of fungitoxicity of 8-quinolinol by amino acids and other chelators. Phytopathology **50**, 421—424 (1960).
ZSOLNAI, J.: Attempts to discover new fungistats. VIII. Antimicrobial activity of new compounds containing an arylazomethyl group. Biochem. Pharmacol. **14**, 1325—1326 (1965).

# Subject Index

# Molecular Biology, Biochemistry and Biophysics

*Volumes published:*

Vol. 1 J. H. VAN'T HOFF: Imagination in Science. Translated into English, with Notes and a General Introduction by G. F. SPRINGER. With 1 portrait. VI, 18 pages. 1967. Soft cover DM 6,60

Vol. 2 K. FREUDENBERG and A. C. NEISH: Constitution and Biosynthesis of Lignin. With 10 figures. IX, 129 pages. 1968. Cloth DM 28,—

Vol. 3 T. ROBINSON: The Biochemistry of Alkaloids. With 37 figures. X, 149 pages. 1968. Cloth DM 39,—

Vol. 4 A. S. SPIRIN and L. P. GAVRILOVA: The Ribosome. With 26 figures. X, 161 pages. 1969. Cloth DM 54,—

Vol. 5 B. JIRGENSONS: Optical Rotatory Dispersion of Proteins and Other Macromolecules. With 65 figures. XI, 166 pages. 1969. Cloth DM 46,—

Vol. 6 F. EGAMI and K. NAKAMURA: Microbial Ribonucleases. With 5 figures. IX, 90 pages. 1969. Cloth DM 28,—

Vol. 8 Protein Sequence Determination. A Sourcebook of Methods and Techniques. Edited by SAUL B. NEEDLEMAN. With 77 figures. XXI, 345 pages. 1970. Cloth DM 84,—

Vol. 9 R. GRUBB: The Genetic Markers of Human Immunoglobulins. With 8 figures. XII, 152 pages. 1970. Cloth DM 42,—

Vol. 10 R. J. LUKENS: Chemistry of Fungicidal Action. With 8 figures. XIII, 136 pages. 1971. Cloth DM 42,—

*Forthcoming volume:*

Vol. 7 F. HAWKES: Nucleic Acids and Cytology. With 35 figures. Approx. 288 pages.

*Volumes being planned:*

AIDA, Kô: Amino Acid Fermentation

ANDO, T., M. YAMASAKI and K. SUZUKI: Protamines: Isolation, Characterization, Structure and Function

ANKER, H. S.: Studies on the Mechanism and Control of Fatty Acid Biosynthesis

ANTONINI, E.: Kinetic Structures and Mechanism of Hemoproteins

BEERMANN, W.: Puffs and their Function in Developmental Biology

BREUER, H., and K. DAHM: Mode of Action of Estrogens

BRILL, A. S.: The Role of Heavy Metals in Biological Oxidation Processes

BURNS, R. C., and R. W. F. HARDY: Nitrogen Fixation in Bacteria and Higher Plants

CALLAN, H. G., and O. HESS: Lampbrush Chromosomes

CHAPMAN, D.: Molecular Biological Aspects of Lipids in Membranes

CHICHESTER, C. O., H. YOKOYAMA and T. O. M. NAKAYAMA: Biosynthesis of Carotenoids

ELEY, D. D.: Biological Materials as Semi-Conductors

FENDER, D. H.: Problems of Visual Perception

GIBSON, I.: Paramecium Aurelia and Cytoplasmic Factors

*Volumes being planned:*

Gorini, L. C., and R. A. Zimmermann: Translation of Genetic Information into Cell Structure and Cell Components

Guillemin, R.: Hypothalamic Releasing Factors

Haugaard, N., and E. Haugaard: Mechanism of Action of Insulin

Heinz, E., and W. Wilbrandt: Biological Transports

Hoffmann-Ostenhof, O.: Biochemistry of the Inositols

Hofmann, Th.: Chymotrypsin and other Serine Proteinases

Jardetzky, O., and G. C. K. Roberts: Nuclear Magnetic Resonance in Molecular Biology

Jollès, P., and A. Paraf: Chemical and Biological Basis of Adjuvants

Kersten, W., H. Kersten and D. Vazquez: Antibiotics Inhibiting Nucleic Acid and Protein Synthesis

Kleinschmidt, A. K., and D. Lang: A Manual for Electron Microscopy of Nucleic Acid Molecules

Klenk, E.: Cerebrosides, Gangliosides and other Glycolipids

Mathews, M. B.: Molecular Evolution of Connective Tissue

Matthaei, J. H.: Protein Biosynthesis

Maurer, P. H.: Chemical and Structural Basis of Antigenicity

McGuire, E. J.: Biosynthesis of Heterosaccharides in Animals

McManus, T. J.: Red Cell Ion Transport and Related Metabolic Processes

Meienhofer, J.: Synthesis of Biologically Active Peptides

Nelson, R. A., and K. Nishioka: The Complement System. Physicochemical Characterization and Biologic Reactivities

Okunuki, K.: Cytochromes

Paul, K. G.: Peroxidases and Catalases

Porath, J.: Modern Methods of Separation of Proteins and other Macromolecules

Quastel, J. H., and C. S. Sung: Chemical Regulatory Processes in the Nervous System

Reeves, P.: Bacteriocins

Rodwell, V. W., and R. Somerville: Metabolism of Amino Acids

Rüegg, J. C.: Introduction to Muscle Physiology

Schanne, O., and Elena Ruiz de Ceretti: Impedance Measurements in Biological Cells

Seliger, H. H.: Bioluminescence

Shulman, S.: Tissue Specificity

Söll, D., and U. L. RajBhandary: Transfer-RNA

Stadtman, E. R.: Enzymatic Regulation of Metabolism

Ts'o, O. P., and P. A. Straat: Nucleic Acids Polymerases

Vinnikov, J. A.: Cytological and Molecular Basis of Sensory Reception

Weissbluth, M.: Physics of Hemoglobin

Zweifach, B. W.: Shock